About IFPRI

The International Food Policy Research Institute (IFPRI®) was established in 1975 to identify and analyze alternative national and international strategies and policies for meeting food needs of the developing world on a sustainable basis, with particular emphasis on low-income countries and on the poorer groups in those countries. While the research effort is geared to the precise objective of contributing to the reduction of hunger and malnutrition, the factors involved are many and wide-ranging, requiring analysis of underlying processes and extending beyond a narrowly defined food sector. The Institute's research program reflects worldwide collaboration with governments and private and public institutions interested in increasing food production and improving the equity of its distribution. Research results are disseminated to policymakers, opinion formers, administrators, policy analysts, researchers, and others concerned with national and international food and agricultural policy.

About IFPRI Research Monographs

IFPRI Research Monographs are well-focused, policy-relevant monographs based on original and innovative research conducted at IFPRI. All manuscripts submitted for publication as IFPRI Research Monographs undergo extensive external and internal reviews. Prior to submission to the Publications Review Committee, each manuscript is circulated informally among the author's colleagues. Upon submission to the Committee, the manuscript is reviewed by an IFPRI reviewer and presented in a formal seminar. Three additional reviewers—at least two external to IFPRI and one from the Committee—are selected to review the manuscript. Reviewers are chosen for their familiarity with the country setting. The Committee provides the author its reaction to the reviewers' comments. After revising as necessary, the author resubmits the manuscript to the Committee with a written response to the reviewers' and Committee's comments. The Committee then makes its recommendations on publication of the manuscript to the Director General of IFPRI. With the Director General's approval, the manuscript becomes part of the IFPRI Research Monograph series. The publication series, under the original name of IFPRI Research Reports, began in 1977.

AF798892

Agricultural Growth and Investment Options for Poverty Reduction in Rwanda

Xinshen Diao, Shenggen Fan, Sam Kanyarukiga,
and Bingxin Yu

RESEARCH MONOGRAPH

IFPRI®

INTERNATIONAL FOOD POLICY
RESEARCH INSTITUTE

sustainable solutions for ending hunger and poverty

Supported by the CGIAR

Copyright © 2010 International Food Policy Research Institute. All rights reserved. Sections of this material may be reproduced for personal and not-for-profit use without the express written permission of but with acknowledgment to IFPRI. To reproduce material contained herein for profit or commercial use requires express written permission. To obtain permission, contact the Communications Division at ifpri-copyright@cgiar.org.

International Food Policy Research Institute
2033 K Street, NW
Washington, D.C. 20006-1002, U.S.A.
Telephone +1-202-862-5600
www.ifpri.org

DOI: 10.2499/9780896291768

Library of Congress Cataloging-in-Publication Data

Agricultural growth and investment options for poverty reduction in Rwanda / Xinshen Diao . . . [et al.].
p. cm. — (IFPRI research monograph)
Includes bibliographical references and index.
ISBN 978-0-89629-176-8 (alk. paper)
1. Agricultural development projects—Rwanda. 2. Agriculture—Economic aspects—Rwanda. 3. Poverty—Rwanda. I. Diao, Xinshen. II. International Food Policy Research Institute. III. Series: IFPRI research monograph.
HD2127.5.A74 2010
338.10967571—dc22 2010003995

Contents

Tables

Figures

Foreword

Successful agricultural development strategies require an understanding of the linkages between alternative agricultural growth options and poverty reduction, if they are to identify appropriate investments in the most effective development programs. The relationships between growth and poverty reduction are not straightforward, however, and there exist trade-offs among different options. This monograph analyzes agricultural growth and investment options for poverty reduction in Rwanda based on an economy-wide approach. It shows that, if an appropriate government investment plan is implemented, the Comprehensive Africa Agriculture Development Programme's (CAADP) target of 6 percent annual growth in agricultural gross domestic product (GDP) is achievable, and the country's poverty rate will be reduced by 25 percentage points over the 1999 rate. However, while the majority of rural households would benefit from rapid agricultural growth, the most vulnerable households—those with very small landholdings and few opportunities to participate in the production of export crops—would likely benefit less.

This report points out that meeting CAADP's 6 percent agricultural-growth target in Rwanda will require the allocation of public resources to the agricultural sector to increase significantly, reaching 10 percent of the total government budget. And estimated economywide returns to public investment in agriculture through linkage and multiplier effects are high: US$1 of investment in agricultural staples generates almost US$4 of increased GDP.

The priority and sequence of investment have been extensively debated among development economists and decisionmakers, and there could be potential trade-offs between rapid sector growth and low economywide returns from various investment arrangements. For example, while the export sector has often attracted more government attention in the form of favorable policies and investment support in many African countries, this monograph shows that such growth does not translate into high economywide returns because of weak linkage effects. These findings call into question whether the export sector is the major driver of poverty reduction. Sharing the favorable policies and investment support enjoyed by the export sector with other agricultural subsectors, such as staples, might prove more beneficial. Governments and donors concerned with African agricultural development would therefore do well to pay greater attention to broad-based development strategies that emphasize inclusive growth.

Shenggen Fan
Director General, IFPRI

Acknowledgments

The authors are deeply thankful to the Ministry of Agriculture and Animal Resources, Rwanda (MINAGRI), in particular Rose Goslinga of the Planning, Policy and Capacity Building Unit of MINAGRI, for sharing the working document *Total Cost of EDPRS Agriculture Sector* (MINAGRI 2008) with us for this project. This document provided detailed targets and costing information, which was crucial for our research. We thank Christian Arnault Emini, from whom we obtained the Rwanda national social accounting matrix (SAM) that he constructed under a joint project with the Ministry of Finance and Economic Planning, Rwanda (MINECOFIN), MINAGRI, and the World Bank. This national SAM provides a foundation for further SAM disaggregation, allowing a computable general equilibrium (CGE) model to be developed for this project. We acknowledge the New Partnership for Africa's Development team in Rwanda and Ousmane Badiane for comments on an earlier draft of this monograph. We also thank Kene Ezemenari, Michael Morris, Liz Drake, and three anonymous external and internal reviewers for their valuable suggestions and Marcia MacNeil for English editing. The research assistance was provided by Valeria Pineiro, who assisted with the SAM disaggregation. Parts of the study were undertaken with the financial support of the Belgian Trust Funds, the U.K. Department for International Development, the World Bank, and the U.S. Agency for International Development.

Acronyms and Abbreviations

ABS	average budget share
AgGDP	agricultural gross domestic product
CAADP	Comprehensive Africa Agriculture Development Programme
CGE	computable general equilibrium
DCGE	dynamic computable general equilibrium
EDPRS	Economic Development and Poverty Reduction Strategy
EICV1	Household Living Conditions Survey, 1999-2001 [Enquete integrale sur les conditions de vie des menages au Rwanda 1]
EICV2	Household Living Conditions Survey, 2005-06 [Enquete integrale sur les conditions de vie des menages au Rwanda 2]
GDP	gross domestic product
IFPRI	International Food Policy Research Institute
MBS	marginal budget share
MDG	Millennium Development Goal
MINAGRI	Ministry of Agriculture and Animal Resources, Rwanda
MINECOFIN	Ministry of Finance and Economic Planning, Rwanda
NGO	nongovernmental organization
PRSP	Poverty Reduction Strategy Paper
RWF	Rwandan franc
SAM	social accounting matrix

Currency

Rwanda	Rwandan franc (RWF)

Summary

Agricultural development strategies that are put forward by individual African countries delineate priorities for actions to enhance agricultural and overall development. Understanding alternative agricultural growth options and their linkages with poverty reduction and prioritizing agricultural investments are the two key components of an agricultural development strategy. However, the relationships between growth and poverty reduction and between targeted growth and required public investment are not straightforward, and solid research is needed to support an evidence-based policymaking process. This monograph provides such a study using Rwanda as a case. An economywide model is developed for the study and is applied to the most recent economic data and public investment information to analyze agricultural growth and investment options for poverty reduction in Rwanda. The monograph shows that the country's targeted agricultural subsector growth, if achieved, would allow Rwanda to meet the Comprehensive Africa Agriculture Development Programme (CAADP) target of 6 percent annual growth in agricultural gross domestic product (GDP) by 2020. With comparable growth in the nonagricultural sector, rapid economic growth would result in the national poverty rate falling to 35.5 percent by 2015, a reduction of 25 percentage points over the 1999 rate. Although the majority of rural households benefit from rapid agricultural growth, the most vulnerable households—those with very small landholdings and with few opportunities to participate in the production of export crops—appear to benefit less. The report shows that economywide growth led by the agricultural sector has a greater effect on poverty reduction than does the same level of growth driven by the nonagricultural sector. Among agricultural subsectors, growth driven mainly by increased productivity in staple crops has the greatest poverty reduction effect.

The report points out that meeting the CAADP 6 percent agricultural growth target in Rwanda will require the allocation of public resources to the agricultural sector to rise significantly and reach 10 percent of the total government budget. Estimated economywide returns to public investment in agriculture are high and will come not only from growth in the agricultural sector. Through linkage and multiplier effects, one dollar of public investment in agricultural staples generates US$3.63 of increased agricultural GDP

(AgGDP) and US$0.21 of increased nonagricultural GDP. In the agricultural sector, economywide returns from investing in staple foods, including staple crops and livestock, are much higher than those from investing in export crops. But even though the investment returns are high, the planned amount of investment in Rwanda will not be enough to significantly improve the current low yields of many foodcrops in the country. The average yield for maize will stay at a low level in 2015—a level already reached by many African countries today.

The report also points out the trade-offs between rapid growth and low economywide returns from investing in the export sector. Targeting the export sector through public policy and investment will bring double-digit growth to the sector, measured by an increase in GDP; however, economywide returns to such investments are low. The weak linkages of the export sector with other economic activities on both the supply and demand sides reduce the role of the export sector as a key driver in both overall economic growth and poverty reduction. Nevertheless, the export sector has often attracted more government attention than has the agricultural sector in many African countries, with favorable policies and investment support. The findings of this report, which show relatively low economywide returns to public spending in the export sector and relatively less poverty reduction from growth led by exports, further emphasize the importance of broad-based agricultural growth. Agricultural development strategy, including effective public investment strategy, has to focus on growth that benefits a majority of farmers. Only such a strategy can be expected to be efficient and effective for growth, poverty reduction, and economic development in general.

CHAPTER 1

Introduction

Rwanda is still marked by the consequences of the 1994 genocide, although the country has made a remarkable transition to peace and development over the past 10 years. Economic policy in this period has focused on reconstruction, and GDP growth averaged 7.3 percent per year between 1995 and 2006. Financed largely by foreign funds through multiple channels of international donors, public investment has picked up and reached an estimated 9.4 percent of GDP in 2007. Security and political stability have been restored, and the business environment has improved; as a result, private investment has increased from about 6 percent in 2001 to an estimated 9 percent of GDP in recent years (Morris et al. 2008). Significant progress has also been made in improving education and health indicators. For example, 5 years after the end of the conflict, the number of primary school students had rebounded to its pre-genocide long-term trend. At 107 percent, Rwanda's gross primary school enrollment ratio is higher today than that observed in other Sub-Saharan countries of similar income level, and the number of students in secondary school has almost tripled since 1996 (Lopez and Wodon 2005). In terms of key health indicators, World Bank (2008a) estimates of infant mortality suggest that after increasing from 85 per thousand to 137 per thousand between 1988–92 and 1992–94, the infant mortality rate receded to 104 per thousand in 1998–2000 and further reduced to 97.5 per thousand in 2006.

Despite the impressive progress that has been made in Rwanda, severe challenges remain for longer term development. The destruction of human and physical capital that ensued from the genocide severely reduced productivity, household income, and government revenue, leading to an increase in poverty. Empirical evidence suggests that the impact of armed conflicts on an economy can be long lasting. According to Lopez and Wodon (2005), although growth rate may recover in a short period after the conflict, because of the loss of all kinds of stock variables, the longer term effect of genocide is a loss to the economy equivalent to 25–30 percent of per capita GDP even many years after the conflict. That is, without the genocide of 1994, Rwanda's per capita GDP could have been 25–30 percent higher in 2001 than it actually was. Indeed, by 2006 per capita income in Rwanda, as measured by the real GDP, is still lower than before the genocide. Challenges also arise from the country's lack of natural resources and its social and institutional constraints. Being both landlocked and resource-poor, Rwanda is often geographically classified into a group of African countries facing the most serious binding constraints in development (Ndulu and O'Connell 2006). In addition, its history of social division and ethnic diversity increases the country's needs for ethnically neutral institutional development (Bigsten and Isaksson 2008). With a per capita annual income of US$260 and about 60 percent of the population living on less than US$1 per day, Rwanda remains very poor and underdeveloped. If Rwanda is to meet its development goals of reducing its poverty rate by one-half and achieving the Millennium Development Goals (MDGs) by 2020, economic growth must accelerate.

Facing such vast development challenges, the government of Rwanda has firmly committed itself to growth and development. The government has set ambitious development goals, and a strategy for reducing poverty and stimulating higher and sustainable economic growth is laid out in *Vision 2020* (MINECOFIN 2000) and is further articulated in the first Poverty Reduction Strategy Paper (PRSP) (MINECOFIN 2002). Agricultural transformation figures prominently in both documents. According to *Vision 2020*, agricultural transformation is expected to boost off-farm growth in both formal and informal sectors, with the effect of reducing the proportion of the population dependent on agriculture from the present 87 percent to about 50 percent by 2020. In addition, because of the low baseline of agricultural productivity, there is considerable potential to achieve substantial income gains by increasing productivity in the short run. In the second PRSP, the Economic Development and Poverty Reduction Strategy (EDPRS) (MINECOFIN 2008), the focus on agriculture (particularly on improving agricultural productivity) and infrastructure has been further strengthened, and the distributional effect of growth is seen to matter more with growth over time (MINECOFIN 2008).

Agriculture features prominently in the Rwandan economy and currently accounts for 40 percent of GDP. Agricultural commodities, mainly tea and coffee, generate 70–90 percent of total export revenues. Modest production gains achieved in recent years in the coffee and tea sectors have allowed Rwanda to broaden its revenue base, but the value per capita of commodity exports remains much lower than the average for Sub-Saharan Africa as a whole. In addition, the overall performance of Rwanda's agricultural sector in recent years has been disappointing. Productivity in many staple crops and the livestock sector has remained flat, while the average farm size has declined. With many rural households surviving from subsistence farming and few growing commercial crops, income growth is stagnant for many farmers. Considering that Rwanda is very densely populated, a traditional subsistence production structure with high population growth has made many farmers' incomes even lower than they were during the pre-genocide period. The goals of increasing market integration of traditional agriculture and creating more opportunities for farmers to produce various agricultural products that can provide more marketable surpluses are always part of a development strategy oriented toward growth (von Braun, de Haen, and Blanken 1991). The key to poverty reduction also lies in stimulating rapid and sustainable growth in the agricultural sector.

The role of agriculture in growth and poverty reduction has been broadly discussed, in particular for low-income African countries (see, for example, Diao et al. [2007] for a synthetic review of the economic literature). Moreover, the recent world food-price crisis further enhances the urgent need to significantly improve the agricultural performance of African countries. A recent World Bank study suggests that the surge in prices could plunge many Africans into poverty (Ivanic and Martin 2008); the effects of food-induced inflation and deteriorating trade deficits on economic growth will only be felt in the years to come. Several studies have also suggested that, unlike the 1974 food crisis, the current crisis may be characterized by higher real food prices for many years in the future (OECD-FAO 2007; USDA 2008; von Braun et al. 2008). In the face of surging food prices, it is especially critical to focus on revitalizing the agricultural sector in African countries for several reasons. In addition to its immediate impact on food security, agriculture is still the largest source of employment in African countries, including Rwanda, and it remains a lead sector of comparative advantage (Diao and Dorosh 2007). Moreover, agricultural productivity growth has repeatedly been shown to be the primary driver of global poverty reduction (Thirtle, Lin, and Piesse 2003; Byerlee, Diao, and Jackson 2005; Christiaensen, Demery, and Kühl 2006; Bezemer and Headey 2008), both through its direct effects on farmers' incomes and through its indirect effect on the reduction of food prices. The sector also has tremendous growth potential when the right policies are in place. Faster agricultural growth has put countries on the path of a much broader transformation process: rising farm incomes raise demand for industrial goods, reduced food prices curb inflation, and induced nonfarm growth increases the demand for unskilled workers. Rising on-farm productivity also encourages broad entrepreneurial activities through diversification into new products, the growth of rural service sectors, the birth of agroprocessing industries, and the exploration of new export markets (World Bank 2008b).

In the context of Rwanda's economy, agriculture is also critical for the country's sustainable development. Even in the 1980s Rwanda was the most densely populated country in Sub-Saharan Africa, with 574 inhabitants per square kilometer of arable land at that time (Clay 1996). Virtually all arable land in the country is used for agriculture. The country also has one of the highest population growth rates in the world: the population doubled in the 20 years between the 1970s and the 1990s. Although 800,000 people (more than 10 percent of the population) died in the genocide of 1994 (Verwimp 2003) and population declined by 23 percent between 1990 and 1995, in 2006 the population was already 34 percent more than its highest pre-genocide level in the early 1990s. Lack of new land for agriculture is also related to the geographic and ecological constraints of the country. The country is dotted with steep hills, where altitudes and slopes change dramatically within shouting distance (Blarel et al. 1992). Land scarcity has compelled farmers to cultivate fragile, steep-slope holdings, which further contributed to the declining agricultural productivity and induced huge environmental costs. A household-level analysis of Clay et al. (1995) shows that on highly eroded farms an additional hectare produces 20–36 percent less than on farms with little erosion. According to the Rwanda Environment Management Authority, without applying such conservation measures as radical terraces, cultivation expanded to steep-slope areas that causes soil erosion has resulted in a loss of 1.4 million metric tons of soil per year. This number represents a decline in the country's capacity to feed 40,000 people annually and is equivalent to 1.9 percent of the country's GDP (REMA 2009). A country like Rwanda urgently needs to emphasize an agricultural strategy that promotes environmentally sustainable land management, invests in soil conservation and fertility, and improves land productivity through various channels.

Against this background, this study assesses the likely effectiveness of alternative agricultural development strategies for stimulating growth and poverty reduction in Rwanda. We expect that this report will provide research-based evidence to support the efforts of the Government of Rwanda to identify areas in which policy reforms supported by public and private investments can best help advance the national agricultural development agenda. Three policy-related questions are the focus of this study: (1) What are the most effective pro-poor options for agricultural growth? (2) How can important linkages between agriculture and nonagriculture be strengthened such that nonfarm activities will become an important source of income for both growth and poverty reduction? (3) What are the most cost-effective public investment choices for stimulating shared growth and poverty reduction? In Chapter 2 of this monograph we introduce the analytic tools and data used for this study. Chapter 3 briefly discusses the structure of Rwandan agriculture and the rural economy. Chapter 4 presents the model results from a "business-as-usual" scenario. Chapter 5 analyzes the agricultural subsectoral contributions to both income growth and poverty reduction. In that chapter we first apply the country's targeted growth to the subsector and crop levels to assess whether these targets can help the country reach 6 percent agricultural annual growth, a goal set by CAADP. We then analyze the relationship between agricultural growth and poverty reduction and assess the feasibility for the country to achieve the MDG of halving its poverty rate in the next 10–15 years. To evaluate the contribution of agricultural growth to poverty reduction, we further analyze the growth at the subsector level and assess which agricultural subsectors are more pro-poor. After clarifying the linkages among agricultural growth, poverty reduction, and growth options among different agricultural sectors in Chapter 5, Chapter 6 focuses on required public investment in agriculture and its priorities to achieve CAADP and poverty reduction goals. In that chapter we first assess the investment required for achieving CAADP growth and poverty reduction MDGs. We then further estimate the returns to public investment at the subsector level and the priorities of investments. Chapter 7 concludes the monograph with major findings that are highly relevant to agricultural development strategies in Rwanda and a discussion of implications for Africa as a whole.

CHAPTER 2

Analytic Tools and Data

Most existing studies in the literature on Rwandan agriculture and agricultural policies are at the micro level with more focus on natural resource management. For example, in a study on farm fragmentation Blarel et al. (1992) used farm-level data of 1987–88 for Rwanda's three prefectures and found both negative and positive aspects of farm fragmentation for farmers. Clay et al. (1995), in collaboration with the Division des Statistiques Agricoles of MINAGRI, measured both the cost of erosion to land productivity and returns to soil conservation investment using the data from national farmhold surveys between 1983 and 1994. Using the data from a similar survey for 1991, Clay (1996) analyzed farmer ability and willingness to invest in conservation and soil-fertility technologies. The study by Roose and Ndayizigiye (1997) analyzed the effectiveness of water and soil-fertility management to fight erosion in the tropical mountains of Rwanda using plot-level data; a study by Kelly et al. (2001a), using the data of 2000's first agricultural season survey, further emphasized the importance of anti-erosion investment and the use of fertilizers to agricultural productivity, rural income, and food security. Beyond agricultural production, Verpoorten and Berlage (2004) conducted a cross-rural household comparison in household strategies to improve income and reduce poverty, whereas Nkeshimana (2008) focused on the relationship between land management practices and soil erosion in the coffee systems of southern Rwanda. The study of von Braun, de Haen, and Blanken (1991) is much more comprehensive and measures the effects of commercialization in Rwanda on production, income, employment, consumption, and nutrition using the household survey data.

Although these studies provide much information and outline the policy options for the country's agricultural and rural economy, no study explicitly focuses on the linkages between this sector and the rest of economy, the contribution of rapid agricultural growth to poverty reduction, and required public investment to support such growth. Furthermore, the methods used in the previous studies are not suitable for analyzing these linkage issues. Thus we adopt an economywide approach to analyze the three broad questions that this monograph addresses. In this chapter we first introduce the analytic tools developed for our study and then briefly discuss the data that was used in the model. Finally, we mention several limitations of the model and our attempts to reduce their effects on the results.

Developing a Dynamic General Equilibrium Model for the Study

An economywide model is the proper tool for analyzing agricultural growth and investment options, as such a model captures synergies and trade-offs from accelerating growth in alternative agricultural subsectors and the economic interlinkages between agriculture and the rest of the economy. For this reason, we have developed a dynamic computable general equilibrium (DCGE) model for this study. A static, standard CGE model was developed in the early 2000s at IFPRI and has been documented in Lofgren, Harris, and Robinson (2001). The recursive dynamic version of the CGE model is based on this standard CGE model with the incorporation of a series of dynamic factors. The early version of this DCGE model can be found in Thurlow (2004), and its recent applications include case studies of three countries: Zambia and Uganda in Diao et al. (2007) and Ghana in Breisinger, Diao, and Thurlow (2009).

Similar to other CGE models, our DCGE model is an economywide, multisectoral model that solves simultaneously and endogenously for a series of economic variables, including commodity and factor prices. However, unlike traditional CGE models that focus on national economies with multiple production sectors, our DCGE model considers subnational heterogeneity in agricultural production by assigning a series of different production functions for producing a similar agricultural product (for example, maize or cassava) to different regions or districts and different types of farms. The setup of such a model requires more information about a country's agricultural production than a traditional CGE model (for instance, information about the distribution of land across regions or districts and types of farms for each individual type of crop or livestock production), which significantly increases the complexity of calibrating the model to the real economy. However, once such information is available and the model is constructed, the model can better capture the economic interlinkages at both the subnational and national levels, including interlinkages across regions and those between sectors. The specifics of the subnational structure of agricultural production used in the model will be further discussed in detail when we introduce the dataset—the Rwanda SAM—and agricultural structure in the next chapter.

Like other CGE models, the DCGE model captures, with its general equilibrium feature, economic activities on both the supply and demand sides. On the supply side, the model has defined production functions for each economic activity, such as agricultural production (for which functions are defined at the subnational level) and nonagricultural production (which is defined only at the national level). As in any other quantitative economic analysis, certain assumptions have to be applied before calibrating the model to the data. In a typical CGE model, a constant return-to-scale technology with constant elasticity of substitution between primary inputs is a necessary assumption for the model to have a general equilibrium solution. However, as both primary and intermediate inputs are considered in the production functions of a CGE model, a Leontief technology with fixed input-output coefficients is often assumed for intermediate inputs, such as fertilizer and seeds in crop production, feed in animal production, and raw materials in food processing, as well as for the relationship between intermediates and primary inputs in aggregation.

The demand side of the CGE model is dominated by a series of consumer demand functions. In our model, the system of consumer demand functions is solved by maximizing a Stone-Geary utility function in which the income elasticity is not necessarily 1 (which differs from a Cobb-Douglas utility function); hence the marginal budget share (MBS) for each consumer good departs from the average budget share (ABS) of this good in the consumer's total budget.[1] With such a utility function assumed, information on income elasticity is required in order to calibrate the demand system to the data. We will discuss this in detail later in the discussion about the data and other parameters applied in the model. As in any other general equilibrium model, consumer income that enters the demand system is an endogenous variable. Income generated from the primary factors employed in the production process is the dominant income source for consumers. Income from abroad (as remittances received) or from the government (as direct transfers) are also considered.

The relationship between supply and demand has to be explicitly modeled in a CGE model and determines the equilibrium prices in the domestic markets. Given that a CGE model also captures trade flows—both import and export—the relationship between domestic and international markets is also

[1] The MBS relates the allocation of incremental income spent on different consumption goods for a consumer, whereas the ABS is the current (total) budget allocation among different goods. For example, suppose that a consumer currently spends 2 percent of her (his) income on rice consumption, indicating that the ABS for rice is 2 percent. When this consumer's income increases in the next year, for each increase of one dollar, she (he) prefers to spend 3 cents on rice. In this case, the value of MBS for rice is 3 percent. When MBS is greater than ABS for a particular consumption good (in this case, rice), demand for this good is called income elastic (Wilhelmsson 2002). However, if MBS is lower than ABS for a particular good, like sorghum, demand for this good (sorghum) is said to be income inelastic. The relationship between MBS and ABS is further discussed in Chapter 3.

modeled explicitly. Generally speaking, any commodity produced or consumed in the domestic market can also be exported or imported. However, in a CGE model, the commodities produced or consumed in the domestic market are not perfectly substitutable for those going to or coming from international markets. Thus the international price for any product, regardless of whether this product is exportable or importable, cannot be fully transmitted into domestic markets, and changes in domestic supply and demand will finally determine its price. However, if a product is exportable or importable, its price in domestic markets can be affected by international prices and by export and import demands. To capture such linkages with international markets, the model assumes price-sensitive substitution (imperfect substitution) between foreign goods and domestic products. Using such an assumption, if domestic demand increases more than the supply of this good, the domestic price for the good rises relative to the export/import prices. Exports of this good fall and imports rise. Conversely, if domestic productivity improves and a rising supply outpaces the increases in demand for the product, the domestic price then falls relative to the border prices, exports rise, and imports fall. Imperfect substitution also implies that agricultural productivity improvement by itself may not be enough to expand agricultural exports: improving marketing conditions is also important.

Although the linkages between demand and supply through changes in income (an endogenous variable) and productivity (often an exogenous variable) are the most important general equilibrium interactions in an economywide model, production linkages also occur across sectors through intermediate demand and competition for primary factors employed in production sectors. Many primary agricultural products need to be processed before reaching consumers and export markets. Food processing is often an important component of the manufacturing sector in developing countries. Productivity-led growth in the agricultural sector can stimulate growth in food processing by providing cheap inputs (forward linkages) and creating more demand for processed goods (backward linkages through rising incomes of farmers). However, growth in an export-oriented agricultural product (for example, coffee or tea in Rwanda) often creates increased demand for processing that product. Although most of such processing activities add very little value, they increase labor demand, and hence create job opportunities for both rural and urban households. Clearly, without a general equilibrium framework and detailed structure for both the agricultural and nonagricultural subsectors, capturing such economywide impacts of agricultural growth is unlikely in a model.

Investments affect production over time, and productivity growth is a gradual process. Capturing this dynamic process is a key component of our

DCGE model. Given the complexity of the model setup for Rwanda—the large number of production sectors in agriculture and nonagriculture, and the disaggregated agricultural production and household groups across subnational regions—it is unrealistic to expect a fully developed intertemporal general equilibrium model for this study.[2] Thus recursive dynamics is applied in the model. With such a model setup, the dynamics takes place between two periods only. In addition, consumption smoothing along the growth path, intertemporal investment, and savings decisions are not taken into account. Instead, private investment (hence capital accumulation) is determined by a Solow-type savings decision in which savings are proportional to income and not endogenously solved from a Ramsey-type intertemporal utility function.[3] Although population (and hence labor supply) growth and land expansion at the subnational and national levels are all exogenously determined, productivity growth in the agricultural sector can be endogenously linked with public investments. In Chapter 5 we link public spending and its allocation to productivity growth in the agricultural sector to assess the returns to agricultural investment.

The government is generally included in a CGE model as an institutional account. In our model, the government collects taxes (including tax revenue from domestic households and producers, export taxes, and import tariffs), transfers part of this income to households, and uses the rest either as investments or recurrent spending. As in many other Sub-Saharan African countries, a major part of government spending in Rwanda is financed by international or developed-country donors. In the model it is captured as a transfer to the government coming from abroad. The required government spending to support agricultural growth is calculated in Chapter 5. Increased government spending can have a certain general equilibrium effect (indirect effect), as it creates more demand for investment goods or services produced domestically (for example, construction services) or imported (capital goods). Change in government demand stemming from increased public investment affects the prices in domestic markets for both commodities and factors. However, given that the public spending to finance public investment is primarily from the international community, we have decided to ignore this more complicated general equilibrium effect. Moreover, including it in the model will not sig-

---

[2] An intertemporal general equilibrium model in the literature is often used with a relatively aggregated economic structure. See Diao, Rattsø, and Stokke (2005) for growth linkage analysis in the case of Thailand.

[3] See Diao, Yeldan, and Roe (1998) for the discussion of Ramsey-type intertemporal utility function and its role in the determination of consumer consumption and savings behaviors.

nificantly change the results. The data and a mathematical description of the DCGE model of Rwanda can be found in Appendixes A and B, respectively.

Microsimulation Module and Poverty Analysis

Although the impact of agricultural growth on the poor can be partially captured directly by the DCGE model simulations, assessing the impact of growth on headcount poverty reduction depends on data that fully capture the poverty distribution of the country. A simple microsimulation module that links with the DCGE model was created for this purpose, based on the Household Living Conditions Survey, 2005–06, or EICV2 (Enquete integrale sur les conditions de vie des menages au Rwanda 2) from MINECOFIN (2007a). Specifically, because household groups in the DCGE model are aggregated from the sample households in EICV2 with their weights defined in the survey, each sample household (with its weight) in EICV2 can be traced to a particular household group defined in the DCGE model. Corresponding to the households and their consumption patterns in the DCGE model, the level and share of consumption for the same commodities by each sample household are also defined using the detailed consumption information included in the survey. Using such information, we create a utility (welfare) function for each sample household in which the levels of consumption (in real terms) for all commodities become variables, whereas the shares of consumption expenditure by commodities are parameters. These micro-level utility functions defined for the sample households (that are the same as those in EICV2)—together with the household, district, and rural/urban identifications—form a microsimulation module that was used for poverty headcount recalculation in each DCGE model simulation.

Change in the sample households' total expenditure[4] over time (measured as change in the utility function) causes some households who are poor to eventually become nonpoor when their total expenditure increases to a level above the poverty line. With some poor households moving from the poor to the nonpoor group, the poverty rate for the country declines over time. To determine which sample household's total expenditure eventually rises to a level above the poverty line, it is necessary to link the change in total expenditure of individual sample households (as a utility function in the microsimulation module) to the change in the DCGE model's household consumption expenditure, which is an endogenous outcome of the model simulation. To do it,

[4] Total expenditure is the measure determining whether a particular sample household is poor or not poor, as the poverty line is defined according to it rather than household income. Household income is less accurate in most living standard surveys than the level of total expenditure.

we have to assume that the level of sample household consumption for each individual commodity changes proportionally to the level of aggregate household group consumption for the same commodity and that this household group in the DCGE model is aggregated from the same sample households as in the microsimulation module. This top-down linkage from consumption of a particular aggregate household in the DCGE model to the consumption of a group of sample households in the microsimulation module allows this module to reflect the differential income and price effects across sample households in determining the new level and pattern of consumption expenditure (and level of welfare). However, because the consumption for sample households in a particular group is proportional to one particular aggregated household that represents the group in the DCGE model, the microsimulation module is unable to capture the income distribution effect within this group. With this caveat in mind, the microsimulation module calculates a new level of total expenditure for all sample households, which allows us to recalculate the poverty headcount for different types of rural and urban household groups as well as for the country as a whole.

Constructing a Highly Disaggregated SAM for Rwanda

The key dataset used in any CGE modeling analysis is a SAM. The 2006 SAM for Rwanda was constructed by Arnault Emini (2007) under a joint project between MINECOFIN in collaboration with MINAGRI and the World Bank. The main part of the SAM construction work was undertaken in Kigali, the capital of Rwanda, by Arnault Emini, with close interaction with the staff from MINECOFIN, MINAGRI, and other government institutions, including the Rwanda Revenue Authority, the National Bank of Rwanda, and the National Institute of Statistics of Rwanda. This SAM was constructed for the national economy, and 51 agricultural and nonagricultural sectors were defined at the national level, with two primary factors—labor and capital—as inputs.

For the purpose of our study, this SAM has to be further disaggregated. Specifically, we separate land from capital account in the national SAM and disaggregate the original sector of "bakery, processed coffee, tea, and sugar" into three subsectors: processing coffee, processing tea, and other food processing. Further SAM disaggregation includes

1. disaggregating agricultural crop production into 30 districts and within each district into two types of farm groups (small farm and medium-large farm);
2. disaggregating livestock production into 30 districts;
3. disaggregating rural households at the district level into two groups (small farmer and medium-large farmer); and

4. disaggregating urban households into two groups (urban households in Kigali and all other urban households).

The SAM disaggregation is motivated by the need to better capture heterogeneity in production structure across districts and source of income across different types of farm groups. Although a majority of Rwandan farmers are involved in staple crop and livestock production, the opportunities to participate in high-value crop production are not equally distributed among them. For example, tea, one of the most important export crops of the country, is only produced in 10 of the 30 districts, mainly in the Western Province. Moreover, tea is produced by large farmers, though small farmers can participate as labor providers. In the case of coffee, the other important export crop of the country, although 19 districts produce coffee, the main production area concentrates in the Southern and Western Provinces. With such uneven distribution in export-crop production opportunities, public support that concentrated on promoting tea and coffee production and exports would likely not benefit those farmers outside the tea- and coffee-growing areas. Without a disaggregated SAM, and hence a disaggregated CGE model, the uneven distribution in income gains from such targeted promotion policies would be ignored, as well as the impact on poverty reduction of these policies.

The disaggregation of the SAM is based on information from EICV2 and the Rwanda Agricultural Survey 2006 (MINECOFIN 2007a, 2007b). The economic structure of the country using this disaggregation will be discussed in the next chapter. After disaggregation, on the production side, there are 960 production activities for 16 crop sectors (16 sectors x 30 districts x 2 types of farms), 270 production activities for 9 livestock sectors (9 sectors x 30 districts), and 28 nonagricultural sectors defined at the national level. On the demand side, there are 62 representative household groups: 60 (30 districts x 2 types of farm households) in the rural area and 2 in the urban area. Given the huge size of the SAM (which is a 1,009 x 1,009 matrix) developed for this study, it is impossible to present it in its entirety in this report. Instead, to help understand the structure of the economy captured by this SAM, we discuss the key structure of the SAM in Chapter 3.

Parameters and Elasticities Applied in the DCGE Model

Any analysis based on a model with a system of equations depends critically on the elasticities and parameters employed in the model. However, unlike most partial equilibrium models (in which supply and demand functions are constructed as elasticity-based functions), in a CGE model, well-behaved structural functions are employed, and they are solved by maximizing profits on the producer side and maximizing welfare on the consumer side. In this way, the parameters capturing the economic structure and factor intensity at the sector level (in our case at the sector and district levels) play more important roles in determining the model results than elasticities do. All these parameters have to be calibrated to the data, together with the predetermined elasticities.

Specifically, the substitution elasticity between primary inputs in the constant elasticity-of-substitution production function has to be assumed or chosen from the literature, because the dataset used to construct a CGE model is generally unable to support an econometric estimation for obtaining such elasticities for the entire production system included in the model. For example, if a Cobb-Douglas technology is chosen as the production structure for a CGE model, it then implicitly assumes a unit elasticity of substitution between primary inputs (such as labor, land, and capital) in the production functions. Thus other parameters in the Cobb-Douglas production function of the model (for example, the marginal product of each input and the key parameter in this type of function) can be directly calibrated using the country data of the SAM (the share of value added for each input employed in the total value added of this sector). In our DCGE model, we chose a general constant elasticity-of-substitution function form (other than a Cobb-Douglas technology) to calibrate other parameters in the production function. The elasticity in the production function is predetermined and is based on the CGE literature on other African countries. The other parameters in the production functions of the model are then calibrated using the data in the Rwanda SAM. In addition, we decided to use a similar substitution elasticity in the production function for each production sector across districts and types of farms. However, because of the difference in factor intensity between different types of farms and different sectoral structures across districts, heterogeneity in the technologies for producing a similar product is captured by calibrating the other parameters of the production function to such disaggregated data.

Besides primary inputs, intermediates are also employed in the production process. With the assumption of Leontief technology in the use of intermediates, there are a set of fixed input-output coefficients applied in the production function, which are directly calibrated using the data of the Rwanda SAM.

Using a Stone-Geary type of utility function, the MBS is the parameter applied in the demand system of the model. Although the ABS for each commodity consumed by each household group can be directly calculated using the data of the Rwanda SAM, to derive a series of MBSs the income elasticity of demand has to be obtained.[5] For this study, the income elasticity is esti-

[5] MBS divided by ABS equals the income elasticity of the demand.

mated from a semilog inverse function suggested by King and Byerlee (1978) and based on data from EICV2 (MINECOFIN 2007a). The estimated results of MBSs for different commodities and different groups of households are reported in Appendix A (see Table A.4 for 1999/2000 and Table A.5 for 2005/06).

Limitations of the DCGE Model

Like any other economic model, the CGE model has its limitations. Of these, there are at least five limitations or caveats that are important to note when interpreting the results. The first caveat is the method used for constructing the dataset (the SAM) for CGE modeling analysis. In contrast to a typical econometric analysis (in which either time series or cross-section data are used to estimate the causal relationship between studied economic and social variables), the dataset (the SAM) used in a CGE model analysis is constructed from one year's data. Given that the agricultural sector is one of the most important economic sectors in many African countries (including Rwanda) and that agricultural production is predominantly influenced by patterns of rainfall and other weather-related factors that often fluctuate over time, the SAM is sensitive to the choice of year. To avoid bias in such choices, it is necessary for CGE modeling researchers to assess a longer period of data for main economic activities, particularly for agricultural production, such that the year chosen for the study is a "normal" year. Although the original SAM for Rwanda was not constructed by us, we have carefully checked whether the year of 2006 is representative for Rwanda's agriculture. The country visits and discussions with local experts have made us comfortable with the year chosen. Moreover, we have looked into the trends for the GDP, AgGDP, and agricultural production for major crops, which further increases our confidence in the use of this year as the base in the study (see Figures A.1 and A.2 in Appendix A for Rwanda's GDP and AgGDP between 1981 and 2006, and Figures A.3 and A.4 for selected crop production between 1998 and 2008).

Although aggregated national economic data can be represented by a time series, the disaggregated household data can only come from an annual household survey, which was not conducted every year (in the case of Rwanda, only two surveys, 1999-2001 and 2005-06, are relevant to our study). Moreover, it is well known that agricultural production data in a typical African country's survey of living standards—the surveys we used in this study for defining household aggregation—are very questionable, because agricultural output is often estimated by the household using recall techniques, which is not further checked by field evaluation (such as conducting crop-cutting measurements using representative farm plots) to obtain independent estimates of crop yields.

This problem can be much more serious for an African country in which root crops and plantain are important components of production and consumption. Although cereal crops are harvested seasonally (once or twice a year), many root and tuber crops, such as cassava and yam or plantain, can be harvested around the year, which makes the recalled output less accurate than that for cereal production. Unfortunately, Rwanda is such a country in which root crops and plantain account for 50 percent of AgGDP. For this reason, it is not proper for us to use the crop production data of the household surveys directly in the model aggregation. Instead, we use the crop area data of the surveys as share to disaggregate the 3 years' average (2005–07) of subnational crop production and area data reported by MINAGRI. Although this method is not optimal for obtaining disaggregate crop data for the model, it at least allows us to avoid serious underreporting for root crops, which makes yield levels more reasonable and results in crop production data consistent with regional and national statistics.

The second caveat of the model has to do with its structure for consumer demand. Even though we have significantly improved the parameter choice in the demand system by econometrically estimating income elasticity of demand and by taking into account subsistence consumption in the demand functions, the use of a linear expenditure system to define the demand system and to specify household demand for individual commodities can only partially capture demand dynamics. MBSs, and hence the income elasticity in such a demand system, remain constant over time. Rapid demand shifts can be better captured by using an implicit direct additive demand system (Yu et al. 2003) or by applying latent separability (Gohin 2005), but the highly disaggregated demand structure in the model constrains our choice of methods. However, given that the current income level is extremely low for a majority of Rwandan households, rapid structural change in household food and nonfood demand will likely not occur over a period of 10 years. Thus this relatively linear demand system may be less questionable for the study in the context of the Rwandan economy. Although the functional form, parameters, and elasticities applied to the demand functions are given, the level of consumption and relative demand and hence consumption shares for individual commodities do change over time with increased income and changes in relative prices, because both income and prices are endogenous variables in the model.

Third, as with most other CGE models, production technologies that are calibrated to the initial economic structure remain constant over time. That is to say, similar to the demand system, production functional forms (including the parameters and elasticities of the functions) are given. That does not imply a constant economic structure over time, as the share of each production sector in the overall economy can change as productivity growth and price evolution vary across sectors and over time. However, with given production functions, the model simulations cannot capture the substantial technological changes and innovations that may be embodied in new investments, especially foreign direct investments, which technically involve changes in the functional forms for those production sectors that are more dynamic. Although this is a shortcoming for a recursive dynamic model, given that the Rwandan economy is at an early stage of its development and foreign investment currently accounts for less than 1 percent of its GDP, significant structural change seems unlikely in this economy over a period of 10 years, which makes this caveat less relevant to our study.

Fourth, in general, the expansion of manufacturing or any sector that is relatively technologically intensive can generate many externalities and spillovers, and the social value of new investments in such sectors can greatly exceed their private value. As such, with its neoclassical theoretical foundation, a typical CGE model is generally unable to capture increasing returns to scale and technological externalities and spillovers. Such models may therefore underestimate the contribution of growth in nontraditional and import-substitutable agriculture and the contribution of new manufacturing activities to structural change if rapid growth occurs over a relatively long period. Although it is impossible to fully overcome the above four caveats, we have run several sensitivity tests whose results are reported in Appendix C.

Finally, the model does not take into account the interaction between agricultural growth and environmental degradation, which has become increasingly important in analyzing agricultural growth options, given changing global climate conditions. As we mentioned in Chapter 1, Rwanda is one of the few African countries in which population pressure and low agricultural productivity have resulted in bringing more and more marginal land into cultivation. Environmental degradation has challenged the long-term development of the Rwandan economy and must be taken into serious consideration in an agricultural development strategy. Although empirical research has firmly established that productivity growth may help in subsistence agriculture, more intensive land use in commercial agriculture (such as export agriculture) to reduce deforestation may have serious long-term consequences for water quality and soil losses. These consequences are exacerbated by tropical soils, which are much more fragile than temperate soils.[6]

By considering natural vegetation cover as a key factor for agricultural production in tropical economies in a dynamic general equilibrium model, Lopez and Niklitschek (1991) have shown that, besides the close relationship

[6] See Angelsen and Kaimowitz (1999) for an extensive review of the relationship among deforestation, economic policies, and development in various economic models.

between population growth and increased pressure on land expansion, there is a more complicated relationship between the relative price for natural resource-intensive agricultural goods and land degradation. Different price policies provide different incentives for the private sector to develop either land- or labor-intensive tropical agricultural production, which can lead to various environmental outcomes. Because our model has not taken this relationship into consideration, it cannot to be used to analyze the effects of different policy options on poverty reduction and income growth when the environmental factor is taken into account. To partially overcome this limitation we discuss possible environmental sustainability impacts in the scenario design and simulations in the following chapters.

Bearing these caveats in mind, the CGE model can provide useful simulations to assess the effects of a particular growth option in the context of a broader economic aspect. Using the parameters and data of the Rwanda SAM discussed in this chapter, the DCGE model is suitable for conducting simulation analyses. Before we discuss the simulations and their results for this study, we review the structure of the economy, particularly the agricultural sector, to better understand the model results.

CHAPTER 3

Agriculture and Recent Agricultural Development Strategies in Rwanda

Agriculture is the most important sector in the economy of Rwanda in terms of its contribution to GDP, employment, and foreign exchange earnings. According to the national account, the agricultural sector currently accounts for 44 percent of GDP, similar to the calculations from the Rwanda SAM used in this study (see Table A.1 in Appendix A for the detailed economic structure in the SAM). However, this figure is likely an underestimate, because it is generally difficult to measure exactly the large amount of food that is produced and consumed by farmers at home. In 2005–06, approximately 90 percent of the economically active population was employed in agriculture. Because many family members work on their own lands, and their incomes, particularly for small farmers, are primarily embodied in the lands they own and farm, the calculated labor income from agricultural production as a share of national total labor income is quite low in the SAM, accounting for only about 30 percent of the total national labor income. Although exports from the primary agricultural sector are small, agricultural exports through processing are large, dominated by tea and coffee exports that account for 31 and 34 percent of national total exports, respectively. In this chapter we briefly discuss some key characteristics of the Rwandan agricultural economy at both national and subnational levels.

Domestic Market-Oriented Agriculture

As a landlocked country with poorly developed roads and other infrastructure both in the country and in neighboring countries, Rwandan agriculture is characterized as a sector with few export opportunities aside from the two traditional export crops, tea and coffee. Approximately 1.4 million rural households depend on agriculture as their main livelihood, and almost two-thirds of their products are destined for their own consumption. According to recent trade data (and also captured in the Rwanda SAM), there are only a very few export crops other than tea and coffee. Moreover, for those export crops, the value of exports is tiny compared to production. One of the most important export crops other than tea and coffee is pyrethrum. As the size of this sector is very small and knowledge of it is constrained by the availability of production data, pyrethrum is aggregated into a subsector called "other export crop group" in the SAM. Exports account for about 4.1 percent of this subsector's output (Table A.1). Other nontraditional exports include vegetables and fruits, but the share of exports in total production is small: 0.62 percent for vegetables and 0.15 percent for fruits. According to Fowler et al. (2007), some staple crops have comparative advantage in regional trade. Regional export-oriented crops include Irish potatoes, rice, wheat, maize, and soybeans. Without information on cross-border informal trade, the SAM captures only a small portion of such trade in Irish potatoes, accounting for 0.40 percent of potato production. Banana production is dominated by cooking bananas but also includes beer and fruit bananas. Some bananas are exported (mainly fruit bananas), but the share of total production that is exported is only 0.01 percent. The share of total livestock production that is exported is also small (3.2 percent). Such exports are identified as exports of "other livestock," meat other than cattle, sheep and goats, and swine and poultry (Table A.1). Lack of market access for exports will become a constraint when growth in agriculture is accelerated. This situation is further analyzed in the model simulation analysis. Table 3.1 lists agricultural commodities and their trade statuses in the SAM and the DCGE model.

Even though about 90 percent of the economically active population is engaged in agricultural activities, some food products must be imported to meet domestic demand in Rwanda. Currently, imports of wheat, maize, and rice account for 20-36 percent of domestic consumption (Table A.1). During the urbanization process, and with population and per capita income growth, imports of these grain products, which are often income-elastic in the early stage of development, are expected to further increase if domestic production cannot catch up with the growth in demand for them. The model simulation analysis addresses this issue in Chapter 5.

Land-Constrained Agriculture Dominated by Rural Households with Extremely Small Landholdings

Rwanda has the highest population density in Africa, rising from 183 per square kilometer in 1981 to 345 per square kilometer in 2000. Rural population per square kilometer of arable land is also one of the highest in Africa (World Bank 2008a). Calculated from EICV1 and EICV2 (MINECOFIN 2003, 2007a), the total agricultural landholding in Rwanda is about 1.1-1.4 million hectares, the average rural household's landholding is 0.75 hectare, and per capita landholding is 0.15 hectare.

Table 3.1 Agricultural commodities in the social accounting matrix and dynamic computable general equilibrium model

Commodity	Exported	Imported
Grains		
Maize	No	Yes
Rice	No	Yes
Sorghum	No	No
Wheat	No	Yes
Roots and tubers		
Cassava	No	No
Irish potatoes	Yes	No
Sweetpotatoes	No	No
Other root crops	No	No
Other crops		
Bananas	Yes	No
Pulses	No	No
Oilseeds	No	No
Vegetables	Yes	No
Fruits (excluding bananas)	Yes	No
Export crops		
Coffee[a]	No	No
Tea[a]	No	No
Other export crops	Yes	Yes
Livestock products		
Beef	No	No
Goat and sheep meat	No	No
Poultry	No	No
Other meat	Yes	No
Fish	No	No
Eggs	No	No
Milk	No	No

Source: Arnault Emini (2007).
[a]Coffee and tea are exported through their processing sectors as defined in the Rwanda social accounting matrix.

Thus farmers with extremely small landholdings are expected to dominate the agricultural economy. As shown in Figure 3.1, including landless rural households, 77 percent of rural households have less than 1 hectare. All these households together own one-third of the total agricultural land in the country, with an average household holding size of 0.37 hectare.

Limited access to land is a key indicator explaining income inequality and poverty in such a country as Rwanda, in which the dominant source of income in the rural area is from the agricultural sector. To capture the correlation between household incomes and landholding size, we ran a regression to

Figure 3.1 Distribution of rural households by landholding size

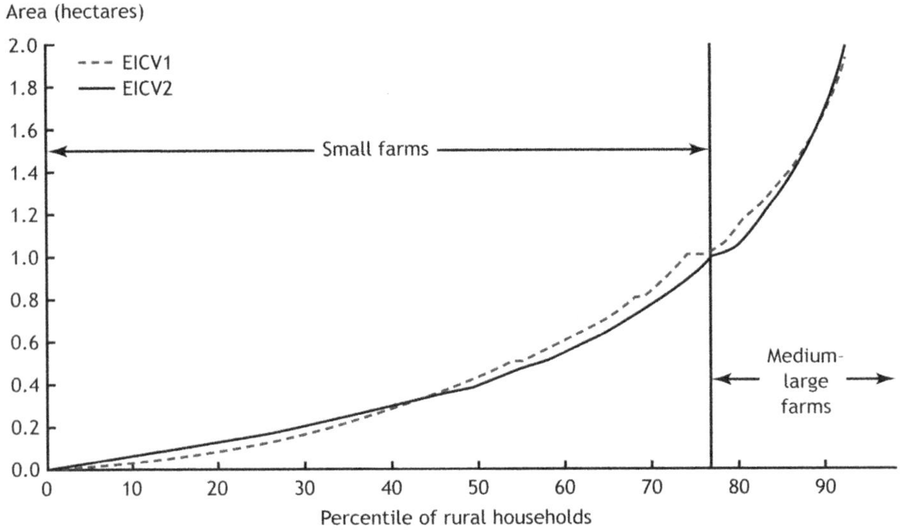

Source: Authors' calculations based on data from MINECOFIN (2003, 2007a).
Note: EICV1 and EICV2, Household Living Conditions Surveys, 1999-2001 and 2005-06, respectively.

formalize this relationship by using a quadratic functional form. As shown in Figure 3.2, there is a strong correlation between the size of landholding and household income, but the relationship is not linear. The *x*-axis of Figure 3.2 represents the average size of landholding for each percentile of rural households, ranked from low to high, and the *y*-axis represents average annual income per capita for the same percentile of households (in thousand Rwandan francs, or RWF). The R^2 value of the regression is 0.64 based on data from EICV1 and 0.47 based on EICV2 (MINECOFIN 2003, 2007a).

The household surveys indicate that, excluding landless rural households, 65 and 61 percent, respectively, of rural households in the country have on average less than 1 hectare of land, whereas 77 and 73 percent, respectively, of the national poor are from these households (Table 3.2). The 1999-2001 national poverty rate was 60.3 percent: 71.1 percent among small farmers and 51.4 percent among other rural households with more than 1 hectare of land. The 2005-06 national poverty rate was calculated as 56.9 percent (National Institute of Statistics 2006). Also, the poverty rate remained highest for small farmers (68.2 percent); it was 46.7 percent for other rural households with more than 1 hectare of land. These results indicate the importance of productivity-led smallholder agricultural growth for poverty reduction in the country. In Chapter 5 we further address this issue.

Figure 3.2 Correlation between landholding size and household income

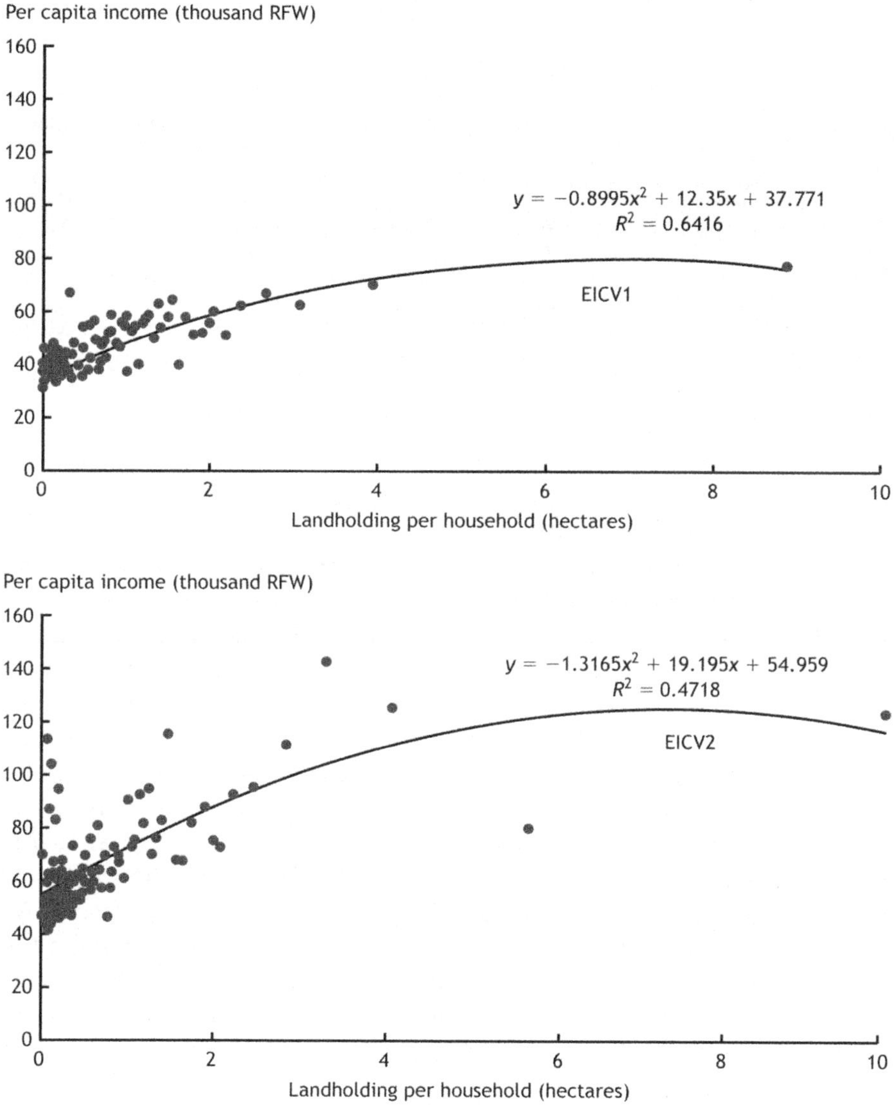

Source: Authors' estimations based on data from MINECOFIN (2003, 2007a).

Notes: EICV1 and EICV2, Household Living Conditions Surveys, 1999-2001 and 2005-06, respectively. The incomes reported for EICV2 are deflated to those for EICV1. RWF, Rwandan franc.

Table 3.2 Distribution of population and the poor

Indicator	Rural			Urban	National
	Holding less than 1 hectare	Holding more than 1 hectare	Total		
EICV1					
Share of population (%)	65.0	24.6	89.5	10.5	100.0
Share of poor population (%)	76.6	20.9	97.5	2.5	100.0
Poverty rate (%)	71.1	51.4	65.7	14.3	60.3
EICV2					
Share of population (%)	61.1	22.3	83.4	16.6	100.0
Share of poor population (%)	73.3	18.3	91.6	8.4	100.0
Poverty rate (%)	68.2	46.7	62.5	28.7	56.9

Source: Authors' calculations based on data from MINECOFIN (2003, 2007a).

Food-Dominated Consumption Patterns

Because of the low income level for the majority of households, food dominates household consumption expenditure, particularly in the rural areas of Rwanda. As shown in Appendix A Tables A.2 and A.3, an average Rwandan rural household spent 77 percent of income on food consumption in the early 2000s (MINECOFIN 2003) and 68 percent in recent years (MINECOFIN 2007a); food accounted for 47 and 43 percent of total consumption spending for an average urban household in these two time periods. Tables A.2 and A.3 also report the consumption patterns of different income groups in both rural and urban areas. We aggregate rural and urban households into 10 groups, 5 in the rural area and 5 in the urban area, according to per capita income and national population quintiles. Not surprisingly, food consumption accounts for a much higher share of total spending among poor households (those in the first two quintiles) in both rural and urban areas. For example, the poorest 20 percent of rural households must spend 86 percent of their incomes on food, whereas for the poorest 20 percent of urban households the food share of total spending is 65 percent.

Measured in real terms, an average Rwandan household's total income (as the level of real total expenditures) increased by 15–16 percent in the 5 years between the two surveys, or about 3 percent annual growth (MINECOFIN 2006),[1] but the share of food consumption in total expenditure only declined

[1] However, until 2006, per capita GDP in Rwanda had still not recovered to its pre-genocide level of 1992–93 (World Bank 2008a).

modestly. This indicates the importance of agricultural development in the country and further justifies the need to increase investment in agriculture to stimulate its growth. The following chapters analyze the impact of agricultural growth through a series of model simulations.

To promote more targeted agricultural growth in a domestic market-oriented agricultural economy such as Rwanda's, it is important to understand whether there exist demand-side constraints among different domestic market commodities. For this reason, we further look into the dynamics in household demand by estimating the marginal propensity of income and calculating the MBS of food commodities based on such estimations. In contrast to the ABS (which is the share of total current income actually spent on each commodity and can be directly obtained from the calculation of the survey data), the MBS refers to the spending patterns of each additional unit of income. Comparing MBS with ABS can increase our understanding of which commodities households would likely prefer to consume as their income increases.

The MBS needs to be econometrically estimated using complete household survey data. In this study, a semilog inverse function suggested by King and Byerlee (1978) was used to estimate the marginal propensity to income. To assess the patterns of marginal demand among different income groups in the rural and urban areas, we report the MBS for each major food commodity by five income groups in the rural and urban areas according to national population quintiles. The following discussion focuses on the estimated results from the most recent survey (MINECOFIN 2007a). Not surprisingly, with the generally low levels of income for most Rwandan households and the domination of food spending in households' total budgets, the MBS of food items in total only declines slightly from the ABS of 59.3 percent to the MBS of 56.7 percent for the country as a whole. That is to say, for each increase of 100 RWF in income, an average Rwandan would spend an additional 56.7 RWF on food and the rest on other kinds of consumption. This situation is consistent with the findings of von Braun, de Haen, and Blanken (1991, 64). Based on the data collected from their survey conducted in 1985–86 in the country, 58 percent of average total household expenditures are spent on basic food items, such as pulses, cereals, roots, and tubers, and the share of total expenditure does not change substantially across expenditure quartile groups in the sample. Although their study uses data from 1985–86, it seems to be very similar to the situation captured in the recent survey for 2005–06 (MINECOFIN 2007a) in terms of the pattern of food consumption expenditure for an average Rwandan household, further indicating the challenge facing the country in promoting agricultural growth and income generation for a majority of households.

Although food as a whole will continue to be the most important spending item for an increased household income, the MBS can be quite different for different types of food. For example, for root crops and bananas combined, the MBS is 37 percent whereas the ABS is 40 percent for rural households. Among different staples, households generally spend more increased income on rice, maize, bananas, and Irish potatoes and less on cassava and sweet-potatoes. In general, with increased income, households will spend more on high-value products, such as vegetables, fruits, oil crops, meat and dairy products, and processed food. However, even though income elasticity for high-value food products is high, staples are still the dominant food items, even for households with increased incomes. For example, staple foods together account for almost 40 percent of each 100 RWF of increased income, whereas all high-value foods account for only 21 percent. To further illustrate the variation of marginal budget spending among different food items, we use two commodities that have a relatively large ABS for rural households: bananas and sweetpotatoes.

As individual commodities, bananas and sweetpotatoes are important staples for rural households, accounting for 6.9 and 9.2 percent of total spending, respectively, for an average rural household. However, poor rural households consume more sweetpotatoes than do their wealthier neighbors, and their share of total spending can be as high as 17.3 percent for the poorest 20 percent of rural households. And yet estimated MBSs show a quite different pattern between these two commodities. Even though their MBS is smaller than their ABS for rural households as whole, the decline in the MBS is much more significant in sweetpotatoes than in bananas. The MBS declines to 2.2 percent from 9.2 percent of ABS in the case of sweetpotatoes; the MBS is 5.0 percent compared with 6.9 percent of ABS in the case of bananas. Moreover, the poorest rural households would like to increase their spending share on bananas using their increased incomes, and they will significantly lower their sweetpotato consumption in this case (Figure 3.3). Obviously, substitution exists between different staples with income growth, which indicates the differences in market opportunities for different staples in domestic markets and must be taken into account in assessing growth and public investment options in the country. In the next chapter, we further analyze such market opportunities based on the model simulations.

Note that declines in the marginal propensity to consume for some staple crops (such as sweetpotatoes and cassava) may create a misunderstanding of the size of the market for each individual product. For example, at the national level, the MBS for sorghum is 40 percent below the ABS (1.0 percent compared with 1.6 percent). One could assume that these numbers imply an absolute decline in national sorghum consumption when per capita income rises. But determining this trend correctly depends on an analysis of the abso-

Figure 3.3 Average budget share (ABS) and marginal budget share (MBS) for rural consumption of bananas and sweetpotatoes

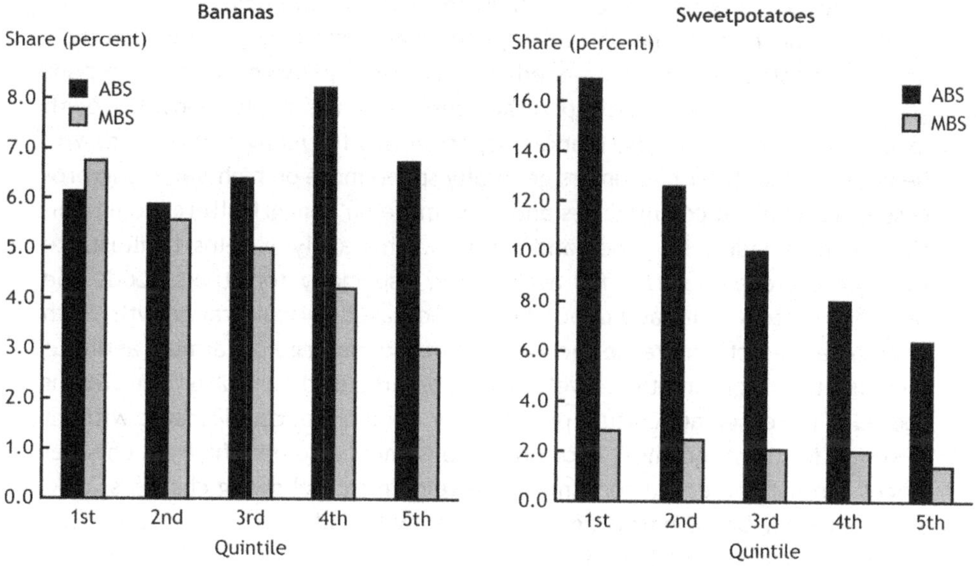

Source: Authors' calculations and estimations based on data from MINECOFIN (2007a).

lute consumption patterns by income groups in addition to spending shares across commodities. According to the 2005–06 EICV2, Rwanda spent 14.2 billion RWF on sorghum consumption, including home consumption by farmers. Processed sorghum products, excluding beer made from sorghum, are also included. Surprisingly, the richest 20 percent of households consumed five times more sorghum in terms of value than the poorest of 20 percent of households, and in total, 36 percent of sorghum available in the country was consumed by the richest 20 percent of the population. A similar situation exists for other staple and root crops, such as maize and cassava, for which the marginal propensity to consume falls with income growth while the absolute amount of consumption continues to rise.

The significant income, and hence expenditure, gap is the key reason for the difference in the absolute value of staple crop consumption between poor and nonpoor households. The total consumption expenditure of an average household among the poorest 20 percent of the population is only one-seventh the level for an average household in the richest 20 percent in rural areas and one-thirteenth in the urban areas. Despite the often smaller ABS and MBS values of food consumption reported for wealthy households, especially for certain staple foods, in absolute terms of food spending wealthy rural households spend 5.2 times as much as poor rural households, and wealthy urban households spend 9.6 times as much as poor urban households.

Both budget share and absolute spending analyses indicate that domestic demand for staples in Rwanda needs to increase rapidly to achieve pro-poor growth and redress the huge gap in the consumption of staple foods. If growth favors wealthy households, market opportunities for many staple foods will be limited. Wealthier consumers generally spend more on high-value and processed agricultural commodities and even more on nonagricultural commodities, such as industrial goods and services. This analysis helps to illustrate that market opportunities for agriculture, especially for staple foods and livestock sectors, critically depend on broad-based agricultural growth. Such growth can directly increase the incomes of the majority of farmers and thus increase their consumption levels. When broad-based agricultural growth is rooted in increased agriculture productivity, food prices can decrease without lowering farmers' incomes. Poor urban consumers also benefit from cheaper prices through increased consumption levels. In the following chapters, linkages between broad-based agricultural growth and poverty reduction are analyzed using our DCGE model.

Patterns of Agricultural Production at the National and Subnational Levels

Rwanda has a diverse diet and hence a diverse agricultural production structure. The final row of Table 3.3 reports the share of each agricultural subsector in the total agricultural production revenue for the country. Although roots and tubers are the largest crop sector in the country, cereals, bananas, pulses, and oilseeds are also important staples in both agricultural production and food consumption. Among root crops, Irish potatoes and sweetpotatoes are the two most important commodities, accounting for more than 27 percent of national agricultural production, followed by cassava, another 7 percent. Bananas are the single most important staple crop, accounting for 17 percent of agricultural production. Among the cereal crops, Rwanda produces maize, rice, sorghum, and wheat. Sorghum is the largest grain crop, followed by maize, but the share of rice in agricultural production (1.7 percent now) has been rising in recent years, driven by increased demand and government supports. Coffee and tea are the most important export crops, but excluding value addition from processing, their value in agricultural production is relatively low, accounting for 2.4 and 1.6 percent of national total agricultural production, respectively. Livestock is less important than crops in agricultural production, consistent with a traditional livestock system constrained by land availability.

Table 3.3 Agricultural production structure at the national and provincial levels (percent)

Region	Commodity							
	Cereals	Roots and tubers	Bananas	Pulses and oilseeds	Vegetables, fruits,[a] and other cash crops	Coffee	Tea	Livestock
Share of subsector's national production								
Kigali	4.1	2.0	2.1	3.3	5.4	0.0	0.0	36.5
Southern Province	20.3	24.3	15.9	23.4	19.1	37.4	21.6	6.4
Western Province	21.7	27.2	18.6	22.5	20.2	41.2	49.3	8.7
Northern Province	18.2	26.1	12.2	19.6	16.1	21.3	29.1	21.1
Eastern Province	35.6	20.4	51.2	31.2	39.1	0.0	0.0	27.3
Share of province's total agricultural production								
Kigali	8.3	16.7	8.5	10.3	17.8	0.0	0.0	38.4
Southern Province	8.5	43.1	13.1	15.0	12.9	4.3	1.6	1.4
Western Province	8.2	43.4	13.8	13.0	12.3	4.3	3.3	1.7
Northern Province	7.9	47.9	10.4	13.0	11.3	2.6	2.2	4.8
Eastern Province	10.2	24.9	28.9	13.7	18.2	0.0	0.0	4.1
Share of total national production	8.8	37.4	17.4	13.5	14.3	2.4	1.6	4.6

Source: Authors' calculations based on data from Arnault Emini (2007).
[a]Excluding bananas.

The structure of subnational-level agriculture diverges significantly from the national average. The first part of Table 3.3 shows the regional distribution of agricultural production (normalizing each subsector's national total to 100), and the second part of the table reports the agricultural structure in each province (normalizing each province's total agricultural production to 100). In general, root crops are equally important for the four provinces except for the city of Kigali. They are the most important staple products in Northern Province, accounting for 48 percent of its total agricultural production. In Eastern Province, bananas seem to be more important, representing 29 percent of its agricultural production and more than 50 percent of the national banana production. The share of cereal production, dominated by sorghum, is also high in Eastern Province: more than 40 percent of the country's sorghum is produced there, accounting for 6 percent of the province's total agricultural production.

Export crops, however, are mainly produced in Western Province. A total of 41 and 49 percent of coffee and tea, respectively, are produced in this province; the remainder is produced in Southern and Northern provinces. There is no coffee and tea production in Eastern Province. Although the share of vegetable and fruit production is the highest in Western Province, the current export opportunities are small for these crops.

As discussed in the previous chapter, the SAM and the DCGE model developed for this study disaggregate the agricultural sector to the district level, with 28 districts that report agricultural production included. Table 3.4 lists the most important districts for producing each crop included in the model. Importance is defined according to the national production share of a specific district. We also report which province each district belongs to in the same table. Sixteen districts appear in each column of the three ranks, totaling 38 district names (some districts appear more than once) in Table 3.4. If a district appears in the table more than once, it indicates that this district is an important producer for more than one type of crop. After dropping all double-counted districts, there are 18 districts that are the most important producers in the country for at least one crop.

The last column of Table 3.4 reports the production share of the top three districts as a percentage of the national total production for each crop. It shows that between 38.4 and 100 percent of the production of export crops (coffee, tea, and other export crops) is concentrated in the top three districts. Rice and wheat production is also relatively concentrated, as the three most important districts produce 48.4 and 58.7 percent of the country's rice and wheat crops, respectively. The next two most concentrated staples are maize (31.6 percent) and potatoes (33.3 percent). For the remaining crops, the top three producing districts account for less than one-third to one-fifth

Table 3.4 Top three production districts for selected crops

Crop	Highest production		Second highest production		Third highest production		Share of national production (%)
	Province	District	Province	District	Province	District	
Grains							
Maize	Western	Rusizi	Northern	Gakenke	Western	Nyabihu	31.6
Rice	Western	Rusizi	Eastern	Nyagatare	Western	Nyamasheke	48.4
Sorghum	Eastern	Nyagatare	Eastern	Bugesera	Eastern	Gatsibo	26.8
Wheat	Southern	Nyaruguru	Southern	Nyamagabe	Northern	Rulindo	58.7
Roots and tubers							
Irish potatoes	Western	Ngororero	Northern	Gicumbi	Northern	Musanze	33.3
Sweetpotatoes	Northern	Gakenke	Southern	Muhanga	Northern	Rulindo	20.8
Cassava	Southern	Muhanga	Southern	Nyanza	Eastern	Gatsibo	20.0
Other roots	Western	Ngororero	Southern	Muhanga	Northern	Gakenke	28.9
Other crops							
Bananas	Eastern	Kirehe	Eastern	Ngoma	Eastern	Nyagatare	29.4
Pulses	Northern	Gakenke	Eastern	Nyagatare	Western	Karongi	20.7
Oilseeds	Eastern	Bugesera	Eastern	Gatsibo	Eastern	Nyagatare	28.3
Vegetables	Eastern	Nyagatare	Eastern	Bugesera	Eastern	Gatsibo	25.3
Fruits (excluding bananas)	Eastern	Bugesera	Northern	Gakenke	Eastern	Nyagatare	23.5
Export crops							
Coffee	Northern	Gakenke	Western	Rusizi	Western	Nyamasheke	38.4
Tea	Northern	Rulindo	Western	Rusizi	Northern	Gicumbi	56.8
Other export crops	Western	Karongi	Northern	Rulindo	Western	Karongi	100.0

Source: Authors' calculations based on data from MINAGRI (2007).

of national production. The concentration of production of export and high-value staple crops in a few districts indicates that the gains from promoting growth in such crops are unlikely to be distributed evenly among districts and rural households. Whether such uneven distribution in high-value production affects income distribution and poverty reduction is an issue relevant to policy. The model simulations discussed in the next chapter help provide an answer to this issue.

Recent Agricultural Development Strategy in Rwanda

With more than 80 percent of the rural population depending on agriculture, agriculture was the only likely candidate for the country's economic development even for previous governments (Ansoms 2008). However, civil conflicts followed by the 1994 genocide did not allow the country to develop and implement a comprehensive agricultural development strategy in that period. Moreover, during the process of reconciliation and rehabilitation in the early post-genocide years, the new government was unable to make agricultural development a top priority. Public expenditure in the agricultural sector was low, and its share in the government's budget declined to less than 3 percent (MINAGRI 2004b).

The first comprehensive strategic document—*Vision 2020*—was published in 2000 as a result of a broad national consultative process that took place in 1998-99 (MINECOFIN 2000). The major aspiration of *Vision 2020* is to transform Rwanda's economy into that of a middle-income country with per capita income of $US900 per year by 2020 (from less than $US300 in 1999). Transformation of agriculture from its current traditional and subsistence system into "a productive, high-value, market-oriented sector with forward linkages to other sectors" was emphasized as one of the six pillars in the documents (MINECOFIN 2000, 3). *Vision 2020* rightly points out that the most important binding constraint of Rwandan agriculture is its low productivity associated with traditional subsistence farming system, not the size of household landholdings. Agricultural policy is to focus on promoting intensification to increase productivity. *Vision 2020* has provided a vision or direction for agricultural development strategy and has outlined eight key policy areas that need urgent attention to bring about the transformation. Being a short document (25 pages), there is no detailed plan in *Vision 2020* for any of these policy areas.

In 2002 Rwanda published its first PRSP (MINECOFIN 2002). Ranked by importance, "rural development and agricultural transformation" moved to the top in a list of the six priority areas, whereas agriculture used to be the second pillar in *Vision 2020* (MINECOFIN 2002, 6). The first pillar was the "reconstruction of the nation and its social capital anchored on good governance, underpinned by a capable state" (MINECOFIN 2000, 3). Moreover, in contrast to the statement in *Vision 2020*, agriculture has been integrated into a broad concept of rural economic transformation in PRSP 2002, and raising agricultural productivity and rural income (including generating opportunities to earn incomes outside agriculture) have become general goals for the transformation (MINECOFIN 2002, 9). Agriculture has been chosen as the primary engine of growth, and PRSP 2002 predicts that agricultural growth will be driven by the increased use of fertilizer (which is targeted to contribute 75 percent of agricultural growth), improvement of wetland management, and crop intensification. PRSP 2002 points out that at the current level of technology, smallholder agriculture is the most productive form of agriculture, and the transformation can be achieved by smallholder households who will be supported by energetic public action. PRSP 2002 also emphasizes the increase and diversification of exports, including agroprocessing exports, to find new engines of growth. The most important development of PRSP 2002 is a detailed discussion of the linkages between agriculture and environmental sustainability. The decline in soil fertility in the country "is compounded by soil erosion and the reduction of the water table in some areas and hence agricultural intensification must be accompanied by environmental actions to manage water flows, control soil erosion and improve the soil structure" (MINECOFIN 2002, 36).

As a part of the action plans outlined in PRSP 2002, a strategy for the agricultural sector was developed through the National Agricultural Policy (MINAGRI 2004a) and the Strategic Plan for Agricultural Transformation in Rwanda (MINAGRI 2004b). The first part of the strategic plan provides an overview of the typology and characteristics of agricultural farms, sector institutions, rural poverty, food and nutrition, land and labor productivity, and the role of women in agriculture. It also presents an analysis that covers strategic aspects of agriculture, including (1) natural resource management and water and soil conservation; (2) crop and animal production and commodity chains; (3) farmer organizations; (4) agribusiness; (5) infrastructure; (6) legal and regulatory frameworks; and (7) financing, coordination, and monitoring of the agricultural sector. The second part of the strategic plan details the strategic plan, priority programs, expected outputs, and contribution of other sectors to support agricultural development; the third part provides a plan of action and estimates the requisite financing. Although 10 strategic aspects of agricultural and rural development are highlighted in the second part of the strategic plan, the document emphasizes that this strategy has to be progressive, flexible, and dynamic.

The strategic plan comprises 4 principal priority programs with 17 subprograms. Their expected results appear in the quantified objectives of the third part of the document as a plan of action. The document also emphasizes the interrelations among the four principal priority programs. These four programs are (1) intensification and development of sustainable production systems, (2) support for the professionalization of producers, (3) promotion of commodity chains and development of agribusiness, and (4) institutional development. Each principal priority program is further composed of subprograms. For each subprogram the document provides a brief synthesis of the diagnosis of the current situation, the strategy to be implemented, specific objectives or targets, and specific actions to be taken. For example, under the first principal priority program (intensification and development of sustainable production systems), there are five subprograms: (1) sustainable management of natural resources and conservation of water and soils; (2) development of integrated agricultural and livestock systems and promotion of specialized intensive animal husbandry; (3) development of marshland and irrigation; (4) supply and utilization of agricultural inputs; and (5) development of methods for establishing food security, managing risk, and assessing vulnerability (MINAGRI 2004b).

The strategic plan also provides a table for requested and committed fund estimations for each program and subprogram (MINAGRI 2004b, 100–103). However, there exists a significant gap between the demands (requested) and supply (committed) in the fund estimation. Although this gap is presented in the table, the document does not discuss how the gap is to be bridged. This shortfall raises the questions of how to prioritize resource allocation to different programs facing such budget constraints (which were already obvious when the plan was developed) and hence, how to implement these programs in such difficult financial circumstances.

CHAPTER 4

A Business-as-Usual Growth Trend: The Baseline Simulation of the DCGE Model

There is no doubt that the Rwanda agricultural strategic plan developed in 2004 (MINAGRI 2004b) is a comprehensive document that is supported by strong evidence developed through a broad consultation process. However, given the constraint of limited financial resources facing the government, how to sequence and prioritize the programs remains a challenge. Moreover, although agriculture is known to be an engine of growth and its growth is generally pro-poor, it still requires empirical evidence to explicitly measure the linkages of growth at the subsector level and the effects on overall economic growth and poverty reduction. The following analysis provides some answers for these highly policy-relevant questions. After introducing the DCGE model in Chapter 2 and briefly discussing the key characteristics of Rwandan agricultural economy and recent agricultural development strategy in Chapter 3, we are ready to apply the model to the Rwandan economy to analyze growth and investment options. To start this analysis, it is necessary to first apply our model to a scenario in which the economy continues its current growth pattern. Thus, in this chapter, we first simulate a scenario of modest growth in both agricultural and nonagricultural sectors to 2015, based on the country's historical data. This scenario is called the "modest poverty reduction in a business-as-usual" scenario.

Because of sharp production declines in 1994 in Rwanda (the year of the genocide) the post-1994 growth rate was comparatively high and has only recently slowed down. According to the World Development Indicator Dataset (World Bank 2008a), for the post-genocide period of 1995 to 2006, annual growth rates for GDP and AgGDP averaged 7.3 and 6.9 percent, respectively, but growth has slowed for both indicators in recent years. Between 2002 and 2006, the annual growth rate for GDP was 4.03 percent, and for AgGDP it was 0.53 percent. The recent poor growth performance in AgGDP reflects the severe drought in 2003 after the above-normal harvest of 2002, followed by another bad year in 2004. Even considering a longer period (for example, between 2000 and 2006), the annual agricultural growth is still low, at 4.1

percent, compared to total GDP growth, which is 5.4 percent in the same period. The data from MINAGRI (2008) show that total crop production grew at 4.7 percent annually between 2001 and 2007. During this period, 30-40 percent of crop production growth resulted from area expansion; the remaining 60-70 percent was due to yield increases (and the majority of the increases represented recovery from the declines of 1994). Such growth is unsustainable, particularly given the land constraint. Consequently, much more modest land-based expansion is assumed in our model, including the promotion of double cropping and intercropping farming practices. Total crop area is assumed to increase by 0.5 percent annually, implying a cumulative increase of about 80,000 hectares of cultivated area (from 1.69 million hectares in 2006 to less than 1.77 million hectares by 2015). The growth rates for individual crop yields is chosen to approximate their national average growth rates from 2001 to 2006, with certain adjustments for some crops with particularly high yields in this period (for example, rice grew at 8 percent, and vegetables and fruits grew at more than 15 percent annually over this period).

The year 2006 was chosen as the base year for the model, which means that the initial yield and area levels by crop used in the model are those reported by MINAGRI for 2006 (MINAGRI 2007). Table 4.1 reports the national levels of cultivated area, yield, and production for the base year by crop. The values of the base year's noncrop production (including livestock, other agriculture, industrial, and service subsectors) are listed in Table 4.2. The model simulation for the baseline run gives the estimated values of the variables in 2015. These variables include area, yield, and production by crop (Table 4.1) and noncrop production value (Table 4.2). The growth rates between 2006 and 2015 are reported in the tables as well.

The DCGE model results indicate that, for a modest growth of 3.8 percent in agricultural production together with 5.1 and 5.7 percent annual growth in industry and services, respectively, GDP grows at 4.8 percent annually and per capita GDP grows at about 2.9 percent. The model result for the growth rate of GDP is higher than that in the period 2002-06 (4.0 percent annually) and slightly lower than in 2001-06 (5.1 percent annually). The simulation results also show a modest reduction in national poverty and improvement in food security. The poverty rate falls to 46.7 percent by 2015, compared to 60.3 percent in 2001 (MINECOFIN 2003) and 57.0 percent in 2006 (MINECOFIN 2007a) although the rate was higher in rural areas (63.4 percent) than in urban ones (28.8 percent). With such a reduction in the poverty rate, together with population growth,[1] the number of the poor would decline by only 220,000

[1] Constrained by the lack of demographic information and population projections at the disaggregated level, we have to assume that the population grows at the same rate for all household groups included in the model.

Table 4.1 Crop area, yield, and production: 2006 estimates and model results from the baseline simulation for 2015, national level

Crop	Area			Yield			Output		
	2006 (thousand hectares)	2015 (thousand hectares)	Annual growth (%)	2006 (metric tons/ hectare)	2015 (metric tons/ hectare)	Annual growth (%)	2006 (thousand metric tons)	2015 (thousand metric tons)	Annual growth (%)
Grains									
Wheat	23	25	1.1	0.9	1.2	3.8	20	30	5.0
Maize	115	126	1.0	0.8	1.0	2.8	92	129	3.9
Rice	14	18	3.0	4.5	4.9	0.9	63	89	4.0
Sorghum	170	162	-0.6	1.1	1.5	3.7	187	246	3.1
Roots and tubers									
Irish potatoes	140	149	0.7	9.2	11.9	2.9	1,286	1,769	3.6
Sweetpotatoes	139	136	-0.2	5.6	6.9	2.4	777	940	2.1
Cassava	119	123	0.4	6.3	8.0	2.8	743	988	3.2
Other roots	25	29	1.7	5.0	6.0	2.2	125	177	3.9
Other crops									
Pulses	388	410	0.6	0.8	1.0	2.7	298	402	3.4
Vegetables	51	51	0.1	5.4	7.3	3.5	271	374	3.6
Bananas	366	393	0.8	7.2	10.0	3.6	2,652	3,927	4.5
Fruits (excluding bananas)	34	32	-0.6	9.9	14.3	4.1	339	463	3.5
Oilseeds	59	61	0.4	0.6	0.8	3.2	36	50	3.6
Export crops									
Coffee	29	34	2.0	0.7	1.1	4.1	21	36	6.2
Tea	13	16	2.0	1.2	1.7	4.1	16	27	6.2
Other export crops	8	9	1.2	19.9	26.5	3.2	162	240	4.5
Total	1,692	1,774	0.5	79.0	104.0	3.1	7,088	9,887	3.8

Source: Numbers for 2006 are authors' calculations based on data from MINAGRI (2007); numbers for 2015 are model simulation results from baseline model at the national level.

Note: Annual growth is projection for 2006-15.

Table 4.2 Noncrop production: 2006 estimates and model results from the baseline simulation for 2015

Subsector	Production		Annual growth (%)
	2006 (million RWF)	2015 (million RWF)	
Livestock			
Cattle	13	19	4.2
Sheep and goats	2	3	4.5
Swine	1	2	3.8
Poultry	1	2	4.6
Raw milk	8	15	7.3
Eggs	1	3	7.5
Other livestock	1	3	7.0
Forestry	39	54	3.8
Fishing	6	9	4.8
Mining	11	16	3.7
Food processing			
Meat, fish, and dairy products	5	8	5.6
Processed cereals	5	9	6.7
Processed coffee	8	17	8.3
Processed tea	9	16	6.7
Bakery and processed sugar	1	2	3.9
Traditional beverages	17	25	4.6
Modern beverages	9	15	5.7
Tobacco	3	5	3.9
Other manufacturing			
Textiles and clothing	5	11	10.3
Wood, paper, and printing	4	5	4.1
Chemicals	7	11	5.5
Nonmetallic minerals	9	13	3.8
Other manufactured products	12	16	3.7
Electricity, gas, and water	9	14	4.5
Construction	97	147	4.8
Services			
Wholesale and retail trade	135	199	4.4
Hotels and restaurants	12	19	5.3
Transport	52	95	7.0
Communications	33	66	8.0
Finance and insurance	74	135	6.9
Real estate	90	166	7.1
Business services	26	49	7.2
Repair	10	18	7.4
Public administration	93	133	4.0
Education	62	91	4.3
Health	20	30	4.4
Other personal services	14	21	4.4

Source: Numbers for 2006 are the authors' calculations based on data from Arnault Emini (2007).
Note: RWF, Rwandan franc.

(from 5.45 million in 2006 to 5.23 million in 2015). The poverty reduction in the model as a result of GDP growth is slightly more optimistic than what was actually experienced in the past 6 years (1999–2005) between the two rounds of household surveys (EICV1 and EICV2). During this period, per capita GDP grew at 2.3 percent annually, while the national poverty rate fell from 60 percent to 57 percent, which indicates that for each 1 percent annual growth in per capita GDP, the national poverty rate fell by 0.42 percent. In the baseline simulation, 1 percent of per capita GDP growth results in a 0.71 percent decline in the national poverty rate. The model results also show that the current gap between supply and demand in the food sector would continue to increase. Imports of maize will double, and rice imports will rise by 70 percent by 2015 from their current levels, making Rwanda more dependent on imports or food aid to meet its basic needs for many staple foods.

CHAPTER 5

Which Sectors Contribute Most to Growth and Poverty Reduction?

Quantitative assessment of how the growth of each agricultural subsector contributes to overall economic growth and poverty reduction is critical for understanding the role of agriculture in development. Recent policy debates on agricultural development in Africa have centered around such issues as the role of smallholders and that of food staples versus export crops. Despite numerous Asian case studies that have proved the important role of agriculture in development, there is doubt about whether agriculture can successfully generate enough growth in Africa today (see, for example, Collier 2002; Maxwell and Slater 2003; Ellis 2005). With rapid expansion in international agricultural trade, many see such high-value commodities as fruits, flowers, and vegetables as the best opportunities for African farmers. Many African countries are being encouraged to expand into high-value, nontraditional exports and to improve the quality of their traditional tree-crop exports. In Rwanda, these sectors, such as coffee and tea, have also attracted more attention and support from the government. Although export agriculture is high value, many such products (including coffee, tea, fruit, and other tree crops) are land intensive. Motivated by the high returns to investment (including land clearing) for these products, land expansion may increase, which may have a long-run negative effect on both sustainable growth and poverty reduction. In this chapter we design a series of subsector agricultural growth scenarios to specifically address the issues related to these different growth options. Based on the actual growth targets at the agricultural subsector level included in the government's document (MINAGRI 2004a, 2004b), we quantitatively assess the following important policy issues:

1. What is the contribution of each agricultural subsector to broad growth and poverty reduction goals?
2. Why is staple growth so important for overall economic growth and poverty reduction?
3. How can staples-led growth help the country meet the CAADP 6 percent growth goal and achieve the first MDG to halve poverty?

4. What is the role of the export agricultural sector?
5. What is the role of nonagricultural growth in this development process?

Given that the model used in the analysis is unable to explicitly consider the joint effect and interaction of environmental degradation and productivity growth, we exogenously control land expansion at the aggregate level in the model. Specifically, we assume that, besides additional irrigated land in rice production and a small amount of additional land for export crops (coffee and tea), additional land will not be brought into cultivation for each year from what has been assumed in the baseline simulation, and thus growth in total area expansion is the same as in the baseline run (0.5 percent of annual expansion at the national level), except in the cases of growth scenarios related to rice, coffee, and tea, which we explain later.

Design of Growth Scenarios to Assess Subsector Contributions to Economywide Growth and Poverty Reduction

To prepare for the country's agricultural strategic plan supporting the new EDPRS (MINECOFIN 2008), targets were set for many agricultural subsectors or specific crops and products. These targets are the result of broad consultation in the country, stock taking of yield and productivity potentials, and comparison with similar neighboring countries in Sub-Saharan Africa. Based on these growth targets, we have designed 21 scenarios to assess each major agricultural subsector's contribution to overall growth and poverty reduction. Growth in most scenarios is driven only by productivity improvement (land expansion is the same as in the baseline run). We present the relationship between scenarios and targeted sectors in Table 5.1; Appendix A Table A.6 lists the assumed exogenous productivity growth rates for each subsector under different scenarios in order to meet the targets set for these subsectors by the government. Only in the scenarios involving growth in rice, coffee, tea, and nontraditional exports is a small amount of new land considered (Appendix A Table A.7). Moreover, under each scenario, the additional annual growth presented in Tables A.6 and A.7 is assumed only for a specific agricultural subsector between 2007 and 2015, while productivity growth in the other subsectors and the rate of total land expansion are maintained at baseline levels.

The government of Rwanda has placed special focus on the promotion of cereal growth, and hence Scenarios 1–4 focus on growth in this subsector by using national targets of the annual growth rate of total factor productivity for maize, wheat, and rice, respectively, in Scenarios 1–3 (see Appendix A Table A.6 for the specific growth targets for these subsectors). Of these cereals, rice has attracted the most public investment. Heightened public invest-

Table 5.1 Model scenarios

Subsector	S1	S2	S3	S4	S20	S21
Maize	x			x	x	x
Wheat		x		x	x	x
Paddy rice			x	x	x	x
Sorghum				x	x	x
	S5	**S6**	**S7**	**S8**	**S20**	**S21**
Irish potatoes	x			x	x	x
Sweetpotatoes		x		x	x	x
Cassava			x	x	x	x
Other roots				x	x	x
	S9	**S10**	**S11**		**S20**	**S21**
Pulses	x				x	x
Bananas		x			x	x
Oilseeds			x		x	x
	S12	**S13**	**S14**	**S15**	**S20**	**S21**
Coffee	x			x		x
Tea		x		x		x
Vegetables			x	x		x
Fruits (except bananas)			x	x		x
Other export crops			x	x		x
	S16	**S17**	**S18**	**S19**	**S20**	**S21**
Poultry	x			x	x	x
Eggs	x			x	x	x
Cattle		x		x	x	x
Sheep and goats		x		x	x	x
Swine		x		x	x	x
Raw milk		x		x	x	x
Other livestock		x		x	x	x
Fishing			x	x		x
Forestry						x

Source: Authors.

Notes: S*nn*, Scenario *nn;* x indicates that the subsector is considered in the scenario. Blank entries indicate that the subsector is not taken into account in the scenario. Scenario 20 considers all staple crops and livestock, and Scenario 21 accounts for all agricultural sectors.

ment in marshland development and the introduction of high-yield varieties have significantly increased rice production area and output in recent years. This strategy is expected to continue for the next 10 years under the national rice development program (Fowler et al. 2007), and hence, the model also assumes additional land expansion for rice production (see Appendix A Table A.7 for details on the annual rate of rice area expansion). Thus, as shown in Table 5.2, the targeted annual growth in rice output is 9.6 percent in the model scenarios.[1]

The production of maize in Rwanda is low compared to its domestic demand. The existing industrial processing capacity of maize is always greater than the domestic supply (Fowler et al. 2007), and imports of maize supply nearly one-quarter of domestic demand (see Appendix A Table A.1). Development of maize production is constrained by its extremely low yields. The national average for maize yields is only 0.8 metric tons per hectare, a level lower than most neighboring African countries in the region (World Bank 2007). Thus in the scenario, maize growth is solely driven by growth in yield while its cultivated area may decline. The targeted maize yield is 1.4 metric tons per hectare by 2015. Although this target is very modest in the case of maize, it still requires an annual growth rate of 6 percent over the next 10 years (Table 5.2).

Besides evaluating the effects on growth and poverty reduction of each grain crop in Scenarios 1–3 when the targeted growth can be met individually in the next 10 years, Scenario 4 combines the first three scenarios together with modest yield improvements in sorghum to simulate joint growth in cereal production.

Scenarios 5–8 are designed to focus on productivity growth in root crops, a subsector that contributes most to the daily diet of the majority of rural and urban households. In recent years, the government has launched a national program to promote special cassava seed production to tackle the mosaic virus

[1] Note that marshland development in Rwanda is controversial in terms of its possible impact on water control and soil erosion. Taking these concerns in consideration, we chose a land expansion rate in rice production of 3 percent annually in the baseline model and 5.8 percent in the growth scenarios (Table A.7). Such area growth rates are much lower than the actual expansion rate in recent years, which was 20 percent between 2001 and 2007 (MINAGRI 2007). However, rice is still a very small crop in Rwanda, and its cultivated areas account for less than 1 percent of total crop area (areas devoted to rice constitute about 4, 17, and 45 percent of those planted in sorghum, maize, and wheat, respectively). The rice yield is much higher than that for the other cereal crops (for instance, rice yields of 4.5–4.8 metric tons per hectare in recent years are 4–5 times those for other cereal crops grown in the country), and there is a substitution effect of rice with other foodcrops. In a general equilibrium model, growth in rice may therefore help to release more land from other cereal crops. Thus the positive impact of rice growth on improving the country's food security and controlling land degradation is probably more than its potential negative effect on the environment.

Table 5.2 Growth in yield and crop production: Model results under the agricultural growth scenario (Scenario 21)

	Yield			Production		
Subsector	2006 (metric tons/ hectare)	2015 (metric tons/ hectare)	Annual growth (%)	2006 (thousand metric tons)	2015 (thousand metric tons)	Annual growth (%)
Grains						
Wheat	0.9	1.4	6.0	20	37	7.2
Maize	0.8	1.4	6.0	92	170	7.0
Rice	4.5	4.7	0.5	63	143	9.6
Sorghum	1.1	1.6	4.4	187	264	3.9
Roots and tubers						
Irish potatoes	9.2	14.7	5.3	1,286	2,211	6.2
Sweetpotatoes	5.6	9.0	5.3	777	966	2.5
Cassava	6.3	10.3	5.7	743	1,151	5.0
Other roots	5.0	13.3	11.6	125	298	10.1
Other crops						
Pulses	0.8	1.1	3.8	298	451	4.7
Vegetables	5.4	7.8	4.2	271	403	4.5
Bananas	7.2	11.0	4.8	2,652	4,642	6.4
Fruits (excluding bananas)	9.9	14.5	4.3	339	515	4.7
Oilseeds	0.6	1.0	5.1	36	62	6.2
Export crops						
Coffee	0.7	1.2	5.9	21	55	11.3
Tea	1.2	1.7	4.2	16	41	11.1
Other export crops	19.9	29.9	4.7	162	354	9.1

Source: Numbers for 2006 are authors' calculations based on data from MINAGRI (2007); numbers for 2015 are model simulation results under Scenario 21 for 2015.

problem that had badly affected cassava production in the country (Fowler et al. 2007). With the continuation of this program, together with a new project to promote the low-cost multiplication of tubers and fruit species, root crops are targeted to grow at 5 percent annually over the next 10 years. Such growth is solely driven by increases in yield, while areas planted in root crops will either remain the same or decline slightly (Table 5.2).

A similar national program has been launched for bananas, the single most important crop for food security in the country. Hence Scenario 10 is designed to focus on banana productivity growth and is used to evaluate its potential contribution to both growth and poverty reduction. Scenarios 9 and 11 are designed to account for the remaining staple crops: pulses and oilseeds.

As we discussed in Chapter 3, agriculture-based products account for 70-80 percent of the country's total export revenues. Thus the two most important export crops, coffee and tea, are always at the top of the government's agenda for investment in agriculture. The national export promotion strategy currently focuses on improving the competitiveness of coffee and tea by increasing their quality and diversification. Through capacity building of farmer organizations and promotion of new technologies, the government expects improvement in the quality of coffee and tea products throughout the production cycle, which will create more export opportunities for these two products. Thus Scenarios 12 and 13 are designed for coffee and tea, respectively. Other nontraditional export crops, such as fruits and vegetables, are considered in Scenario 14. Scenario 15 combines Scenarios 12, 13, and 14 to capture potential growth in both traditional export and nontraditional export commodities.[2]

Compared to crop production, the livestock subsector is relatively small in the Rwandan agricultural economy. The small size of this sector in the economy is mainly due to the high proportion of local species that have low productivity and the generally poor standards of animal husbandry (Fowler et al. 2007). However, demand for livestock products is expected to grow faster than for many other staple foodcrops because of high income elasticities for them. Many different government projects—together with some sponsored by nongovernmental organizations (NGOs)—have been launched recently to promote livestock production, and most are expected to continue over the next 10 years. Thus Scenarios 16–19 are designed to evaluate the effects on growth and poverty of growth in livestock and fisheries. Specifically, Scenario 16 focuses on poultry and egg production; Scenario 17 addresses cattle, sheep and goat, swine, and milk production; Scenario 18 focuses on fish production; and Scenario 19 looks at total livestock plus fish production.

After evaluating the potential contributions of individual agricultural subsectors to overall economic growth and poverty reduction in Scenarios 1–19, Scenario 20 is designed to simulate the joint effect of growth in all agricultural staples (both crops and livestock). In other words, this scenario is the combination of Scenarios 1–3, 5–7, 9–11, and 16–18. The purpose of this scenario is to quantitatively assess the importance of staple-led growth in improving the country's food security, overall economic growth, and poverty reduction. Scenario 21 combines growth in staple crops and livestock with growth in export crops. Using this scenario, we can assess whether the CAADP 6 percent growth target for the agricultural sector can be supported by the growth targets for individual agricultural products developed by the govern-

[2] Although improving the competitiveness of export crops is government policy for promoting growth in export crops, given that many export crops are land intensive and demand for them is barely constrained by domestic consumption (which is often the case for staple foodcrops), the possible impact of rapid growth in export crops on deforestation and land degradation should be taken into account, particularly in promoting large-scale, commercialized production (such as for tea). However, we are unable to measure such impacts in this study.

ment. This scenario can also be used to help determine whether the country will be able to meet the first MDG by 2015, assuming 6 percent annual agricultural growth.

Finally, a scenario that is not listed in Table 5.1 (Scenario 22) considers comparable growth in the nonagricultural sectors in addition to growth in the agricultural sector. This scenario is used to evaluate the linkage and synergy effects of such growth and how it will help the country achieve its broad development goals and the first MDG.

We must point out that national growth projections for agricultural products in government documents often take the form of targeted production. Considering the actual constraint of land expansion in the country, we focus on yield to simulate such growth in the model. In other words, we convert targeted production into average annual growth of total factor productivity in 2007–15, based on the level in 2006, whereas additional growth by means of land expansion is considered only for rice, coffee, tea, and nontraditional export crops. Additional growth in total factor productivity is an exogenous factor that is assumed in the simulations, whereas the simulation results for crop yields and crop production (and hence their growth) are endogenous variables. Allocation of factors (labor, land, and capital) across subsectors of agriculture as well as the use of intermediate inputs, which are endogenous variables in the model, respond not only to growth in total factor productivity but also to changes in relative prices for agricultural and nonagricultural products. We report simulation results for crop growth under Scenario 21 (the scenario simulating growth in all agricultural sectors) in Table 5.2; growth for noncrop production in the same simulation is listed in Table 5.3.

The 6 Percent CAADP Growth Target Is Reachable

Questions regarding whether the government's detailed production targets for the agricultural subsectors can support the goal of 6 percent total agricultural growth and how such growth contributes to the realization of the first MDG remain to be answered. Scenario 21 is designed to model the joint effects of growth across agricultural subsectors. The model shows that if the desired targets at the subsectoral level can be achieved, AgGDP would grow 6.3 percent annually between 2007 and 2015—almost doubling baseline growth. This result represents total GDP growth of 6 percent, compared with the 4.8 percent baseline. If additional yearly growth of 2.8 and 2.4 percent, respectively, is assumed for the industrial and service sectors—comparable with government targets (Scenario 22)—interlinkages between nonagriculture and agriculture fuel AgGDP growth to 6.5 percent per year, and total GDP growth rises to 7.4 percent per year. At this rate, per capita GDP grows 5.5 percent annually, doubling the baseline level. With such high growth rates,

Table 5.3 Growth in noncrop production: Model results under the agricultural growth scenario (Scenario 21)

	Production		
Subsector	2006 (million RWF)	2015 (million RWF)	Annual growth (%)
Livestock			
Cattle	13.4	24.9	7.2
Sheep and goats	1.8	3.7	8.1
Swine	1.2	2.3	7.4
Poultry	1.4	3.2	9.3
Raw milk	7.8	27.5	15.0
Eggs	1.3	4.6	14.9
Other livestock	1.5	7.5	19.6
Forestry	38.9	79.2	8.2
Fishing	6.0	11.3	7.3

Source: Numbers for 2006 are authors' calculations based on data from MINAGRI (2007); numbers for 2015 are results from model simulation under Scenario 21.
Note: RWF, Rwandan franc.

national poverty by 2015 falls to 35.5 percent—24.5 percentage points lower than the rate in 1999 and 11.2 percentage points lower than the 2015 baseline. If such growth trends continue, the country would achieve the first MDG of halving the 1999–2001 poverty rate of 60 percent by 2020.

Effects on Differential Income and Poverty Reduction

Growth may not equally benefit rural households. Empirical studies in other countries often show that rapid economic growth does not always result in shared growth (Akita and Kawamura 2002; Kanbur and Zhang 2005). Differences in poverty reduction and income growth across regions have also been observed in China (Chen and Ravallion 2000). Thus it is essential to further assess the effects on income and poverty of agricultural growth at the household level.

In the case of Rwanda, the simulations indicate that rapid agricultural growth benefits the majority of rural households, and that the distribution of benefits is relatively equal. Nevertheless, the household group with the smallest landholding appears to benefit less than the group with medium and large landholdings. Under Scenario 22 (in which both the agricultural and nonagricultural sectors follow their growth targets) annual income growth for the small-farm group at the national level is 8.47 percent, whereas it is 9.03 percent for the household group with larger landholdings (Table 5.4, second column). Rural poverty falls across all provinces and household groups, but at the national level, poverty falls more among the larger landholding group

Table 5.4 Rural income growth and poverty reduction: Model results under the total growth scenario (Scenario 22), by farm group (percent)

Farm size/region	Income growth rate			Poverty rate		
	Baseline growth	Growth rate under Scenario 22	Additional growth from baseline	2006	By 2015 in Scenario 22	Poverty reduction
Rural small farm households						
Kigali	5.27	7.86	2.59	50.4	30.8	-19.6
Southern Province	5.91	8.44	2.53	72.7	47.4	-25.3
Western Province	6.09	8.82	2.73	65.6	40.2	-25.4
Northern Province	5.78	8.37	2.58	66.8	39.6	-27.2
Eastern Province	5.43	8.09	2.65	54.0	31.0	-23.0
National	5.83	8.47	2.63	64.9	39.9	-25.0
Rural medium-large farm households						
Kigali	5.05	7.60	2.56	50.4	32.9	-17.5
Southern Province	6.98	8.97	1.99	70.2	42.5	-27.7
Western Province	7.47	9.77	2.31	63.2	39.9	-23.3
Northern Province	7.06	9.35	2.29	58.8	33.5	-25.3
Eastern Province	5.77	8.43	2.66	48.7	24.1	-24.6
National	6.65	9.03	2.38	58.4	33.1	-25.2

Sources: Income growth rate and poverty rate are model simulation results under the baseline and total growth scenario (Scenario 22). Poverty rates in 2006 are estimated based on EICV2 (National Institute of Statistics 2006).

than among the small farm group (Table 5.4, last column). As the initial poverty rate is already higher among small farmers, the difference in the poverty rate between these two types of rural households is further widened by 2015. At the provincial level, the poverty rate is highest in Southern Province for both household groups. Although the initial poverty rate of the Southern small farm group is 2.5 percentage points higher than for the other Southern farm group, the poverty gap between these two groups increases to 4.9 percentage points by 2015, although the poverty rate falls in both groups.

Growth in Staple Crops Is More Pro-Poor Than Other Agricultural Growth

The variation in effects on poverty reduction of growth at the provincial level and among the two types of farm households relates to the different income sources of these farmers and the local agricultural structure they face. Because differing linkages between poverty and growth could occur in the agricultural subsector, it is important to understand such linkages to provide insight into designing a pro-poor growth strategy. For this purpose, the poverty-growth elasticity was calculated to enable direct comparison of the various outcomes for poverty reduction.[3] This elasticity was calculated for the two broad agricultural categories—staple food and export crops—as well as for each individual crop or livestock product for which targeted growth was individually simulated in Scenarios 1-18. The poverty-growth elasticities are endogenous outcomes of the model results. Growth affects individual households differently because of heterogeneity across household groups. As shown in the above analysis, with different income sources and land sizes, growth in income and reduction in poverty across household groups vary considerably from average changes at the national level. To capture linkages between growth and poverty, changes in the distribution of incomes, which are primarily determined by the initial conditions used in the model for the country, need to be understood. For example, households with greater opportunities to produce higher value export agricultural products may be better positioned to benefit from export agriculture. But because most export crops can be grown only in certain areas of the country—where households involved in export crops are usually less remote and less poor—economic growth driven by agricultural exports may in fact have less impact on poverty than growth driven by staple production. In contrast, staple crops are often a more important source of agricultural incomes than export crops in the poorer (and more remote) regions of the country. Hence, because staples have a greater impact as an income source for the poor, growth in the production of staple crops is expected to be more pro-poor than growth in the production of agricultural exports.

Although agricultural growth is generally pro-poor, a gap in the poverty-growth elasticities between staple crops (such as grains and roots) and agricultural exports indicates the importance of staples for poorer rural households (Table 5.5). If economywide growth is led by growth in grain crops (Scenario 4), each 1 percent of additional growth in per capita GDP leads to an addi-

[3] The poverty-growth elasticity used in this study measures the responsiveness of the poverty rate to changes in the per capita GDP growth rate. The formula for this elasticity is

$$\frac{\Delta P_0 / P_0}{\Delta GDP_{pc} / GDP_{pc}} = \frac{\Delta P_0}{\Delta GDP_{pc}} \times \frac{GDP_{pc}}{P_0},$$

where ΔP_0 and ΔGDP_{pc} are average annual changes (from the base year) in the poverty headcount rate and level of per capita GDP, respectively; and P_0 and GDP_{pc} are the base-year poverty headcount rate and per capita GDP, respectively. The poverty-growth elasticity measures the percentage change in the poverty headcount rate caused by a 1 percent increase in per capita GDP. This value is not equivalent to a percentage point change in the poverty headcount rate.

Table 5.5 Poverty-growth elasticities: Model results

Subsector	Elasticity
Grains	
Maize	-2.39
Rice	-1.86
Wheat	-1.60
Roots and tubers	
Cassava	-1.60
Irish potatoes	-1.40
Sweetpotatoes	-1.65
Other crops	
Pulses	-2.59
Bananas	-2.05
Oilseeds	-2.17
Export crops	
Coffee	-1.81
Tea	-1.63
Other export crops	-2.27
Livestock	
Poultry	-0.45
Other livestock	-1.38
Fishing	-2.11
Grains	-1.74
Root crops	-1.54
Livestock	-1.35
Export crops	-1.68
Agriculture	-1.53
Agriculture with transport	-1.37
Nonagriculture	-0.49
All sectors	-0.97

Source: Numbers are model simulation results derived from authors' calculations using 2006 data from MINAGRI (2007).

Note: The elasticity value indicates the resultant percentage reduction in the national poverty rate for 1 percent of additional growth in the per capita gross domestic product.

tional 1.74 percent decline in the national poverty rate. This reduction effect is particularly large in the case of maize, with an elasticity of 2.39. However, if economywide growth is led by growth in export crops—including both traditional export crops, such as coffee and tea, and nontraditional ones, such as vegetables and fruits (Scenario 15)—1 percent additional growth in per capita GDP reduces the national poverty rate by 1.68 (Table 5.5).

The model results indicate that putting staples at the top of the agenda can promote broader economic progress and poverty reduction in Rwanda.

This conclusion is in general valid for many African countries, as smallholders constitute more than 70 percent of the continent's farmers (Johnson, Hazell, and Gulati 2003). A wide range of research has also demonstrated the importance of food staples in driving growth and contributing to a dynamic structural transformation of rural economies (Byerlee, Diao, and Jackson 2005; Hazell and Diao 2005; Bezemer and Headey 2008; World Bank 2008b). Acceleration in staples production has also been found to produce secondary and tertiary effects on the broader economy by reducing food prices for urban consumers, curbing overall inflation, and releasing scarce foreign exchange for the importation of goods that are unsuited to production in Africa (Diao et al. 2007). And in the longer run the productivity growth in staples agriculture will facilitate a more fundamental transformation in the broader economy through new opportunities for industry (such as agroprocessing), growth opportunities for rural nonfarm activities (Haggblade, Hazell, and Reardon 2007), increased regional and international trade, and new employment options through expanded migration.

Growth Effects on Agricultural Prices and Trade

Growth may not always benefit producers, especially when it is unbalanced and occurs only in a few agricultural subsectors. When growth targets are set too high for some agricultural production, there is not enough demand from domestic markets, and if it is difficult to export the resultant increased supply, prices can fall significantly. Price drops may hurt some farmers if they cannot adopt more efficient technologies in their production process. For this reason, it is necessary to look at the effects of growth on prices, particularly prices for the agricultural products widely grown by small farmers. Here we focus on Scenario 22 (in which both agricultural and nonagricultural growth is accelerated) and have selected seven crop and livestock commodities to discuss price effects.

As shown in Figure 5.1, prices for most staple crops will either not decline or decline modestly if agricultural and nonagricultural sectors grow simultaneously. This result is consistent with the data from the two surveys of household living conditions discussed in Chapter 3 (MINECOFIN 2003, 2007a). Given the current extremely low level of income (and hence consumption) in Rwanda, the demand for staple crops increases with income growth if broadly based growth can bring more income to the majority of rural and urban households, particularly poor households. The domestic market will have to become the dominant destination to absorb the increased supply for most staple crops. In fact, imports of maize have actually grown quite rapidly, driven by increased demands for food and feed, indicating additional room for further growth in maize production. Similarly, the domestic price for rice

Figure 5.1 Changes in price for four staple commodities: Model results under the scenario for agricultural and nonagricultural growth (Scenario 22), 2006-15

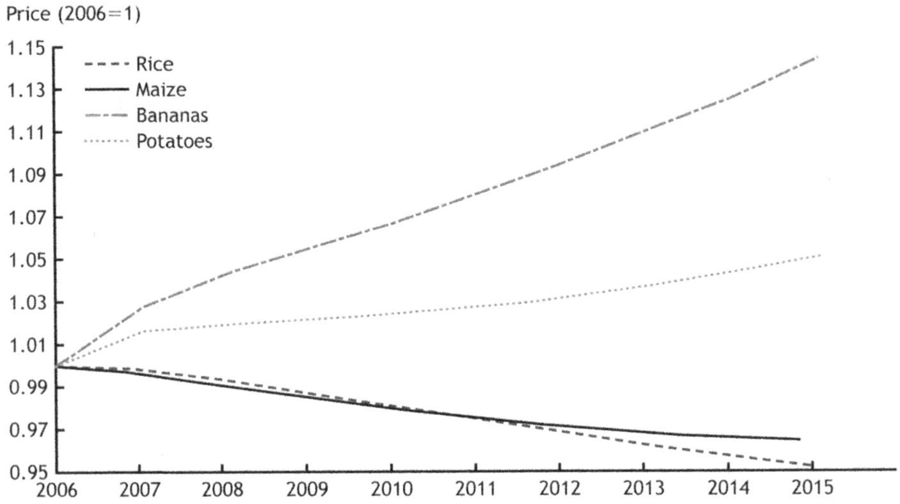

Source: Numbers for 2006 are authors' calculations based on data from MINAGRI (2007); numbers for subsequent years are model simulation results under Scenario 22.

Notes: Prices for 2006 are normalized to 1.

Figure 5.2 Changes in price for selected livestock products and vegetables: Model results under the scenario for agricultural and nonagricultural growth (Scenario 22), 2006-15

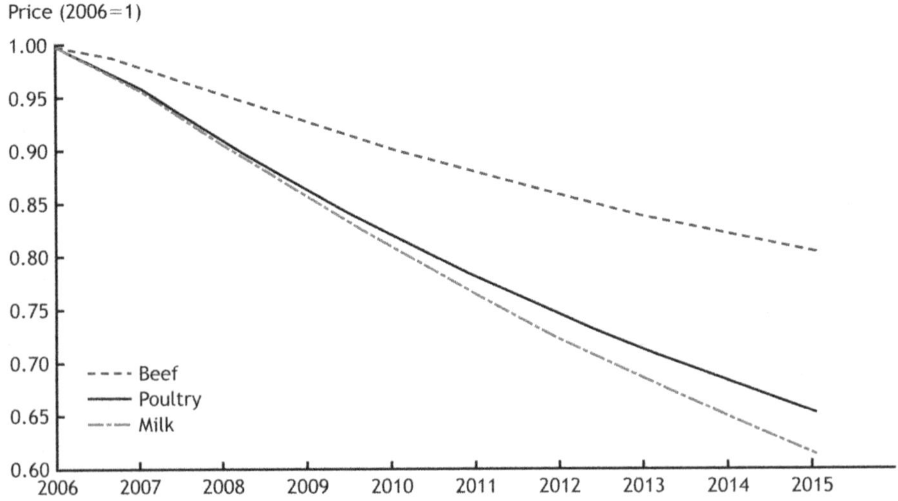

Source: Numbers for 2006 are authors' calculations based on data from MINAGRI (2007); numbers for subsequent years are model simulation results under Scenario 22.

Note: Prices for 2006 are normalized to 1.

is expected to fall modestly, causing import substitution. Although domestic demand for rice will double over 2007–15, the ratio of imports to domestic consumption will fall from its current value of 40 percent to less than 30 percent by 2015. It is also important for the country to reduce its dependence on imported staple food and hence to improve its food security.

We display the changes in prices for selected livestock products in Figure 5.2. As the growth targets are very high for these commodities, it is not surprising to see that, although their income elasticities are high, their prices fall dramatically. Poultry and milk prices fall by 35–40 percent in 9 years. Starting from a very small base, poultry and milk production and consumption will have double-digit growth rates in this period. Although declining poultry and milk prices benefit consumers, the low prices may push some small farmers out of production, unless they significantly improve their productivity.

As discussed in Chapter 3, Rwanda depends heavily on agriculture for export revenues. Agricultural exports account for 90 percent of total exports. This reliance on agricultural commodities for export earnings will deepen with rapid growth in export sectors. Under Scenario 22, growth in coffee and tea reaches

Figure 5.3 Growth in agricultural and nonagricultural trade: Model results under the scenario for agricultural and nonagricultural growth (Scenario 22)

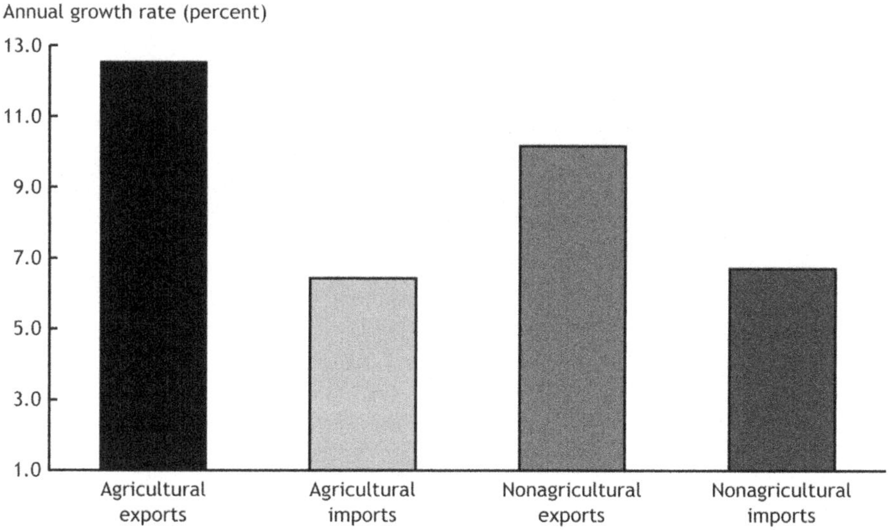

Source: Numbers for 2006 are authors' calculations based on data from MINAGRI (2007); numbers for subsequent years are model simulation results under Scenario 22.

Note: The ratio of net agricultural exports to net nonagricultural imports is 0.08 and 0.29, respectively, for 2006 and 2015.

16 and 14 percent annually, respectively, and exports of nontraditional agricultural products grow at 28 percent annually. These rates result in the growth of total agricultural exports by 13 percent each year (Figure 5.3). However, agricultural imports grow much more slowly, at less than 7 percent, because of import substitution in many agricultural food commodities.

Although nonagricultural exports grow more rapidly than nonagricultural imports, the quantity of nonagricultural exports is much smaller than that of imports. This imbalance results in an increased deficit of nonagricultural trade. Thus the agricultural trade surplus becomes increasingly important to finance nonagricultural imports. We calculate a ratio of agricultural trade surplus over nonagricultural trade deficit, and, as shown in Figure 5.3, this ratio rises from its current of about 0.08 to 0.29 by 2015 when both agricultural and nonagricultural sectors are allowed to grow simultaneously under Scenario 22.

CHAPTER 6

Agricultural Spending Required to Achieve CAADP and Poverty Reduction Goals

Achieving the growth in Rwanda's agricultural sector required to meet both CAADP and poverty reduction goals must be supported by large amounts of investment from both the private and public sectors to improve farming and business conditions and to bring modern technology and knowledge to farmers. The private sector has been involved in such investments in many countries outside Africa, but at Rwanda's early stage of development—and its concomitant extremely low income and savings rate for a majority of the rural population—the private sector is unlikely to play a leading role in such investment. Moreover, investments in rural and agricultural development (such as terracing, irrigation, rural road construction, and other infrastructure) and agricultural research and extension have aspects of public goods, whose development is often the responsibility of the government. Thus public investment is an important instrument not only in improving the quality and provision of public services but also in attracting private investment and inputs. The discussion here focuses on public-sector spending in agriculture required to achieve these goals and the potential returns to investments in different subsectors of agriculture.

The analysis in Chapter 5 indicates that AgGDP could grow at more than 6 percent annually over the projection period 2007–15 if agricultural commodity or subsector growth reaches the national targets set by the government. These growth targets are consistent with CAADP goals and will significantly reduce poverty. The model results show that with more than 6 percent agricultural growth and rapid growth in the nonagricultural sector, Rwanda will be able to achieve the first MDG of halving poverty by 2020.

To promote more rapid agricultural growth and greater poverty reduction, the Government of Rwanda has already committed to increasing its investment in agriculture, and many agricultural development programs are being implemented. For example, three types of investment programs are currently being implemented for marshland development, among which the national rice development program is estimated to be valued at about 330 million RWF.

Many development strategies targeting sectoral productivity, production capacity, commodity quality, and competitiveness are also under way. Moreover, the government has also increased investment in rural infrastructure, markets, and supply chains to improve the external environment for agricultural growth and rural development. Such large-scale public investment is not only necessary for agricultural growth and rural development but is also a precondition for the private sector, including farmers, to increase investment. At this early development stage, marked by low incomes for most rural households, the savings rate of Rwandan households is extremely low. Farmers have no financial capacity to invest in agriculture-related infrastructure, such as marshland development and terracing. However, when the government increases its investment in these areas, the labor inputs are often provided by farmers through food-for-work or other types of government projects. Once terracing and marshland development have taken place, the incentives for farmers to maintain and further improve the lands by additional labor inputs will increase, because farmers know that such inputs can provide them with sustainable higher yields for crops and hence with higher returns. Farmers will also have increased incentives to use modern inputs, such as fertilizer, on their terraced farmlands, as returns to such inputs are higher on such improved lands than on lands without any investment.

These interventions and investments will build a solid foundation for higher agricultural growth in the future, but the short implementation period makes the ex post assessment of the impacts of these endeavors on future growth difficult. For this reason, an ex ante approach is developed for this report. In the first two sections of this chapter, we focus on the amount of public investment in agriculture required for achieving the growth targets discussed in Chapter 5. In the rest of this chapter, we introduce the newly developed ex ante approach and apply it to assess the potential returns of agricultural investment and the relationship between targeted growth and increased agricultural investment at the subsector level.

Current Agricultural Spending Trends

Published data in *Annual Finance Laws, 1999-2006* (MINECOFIN 2006) show that the share of public resources allocated to the agricultural sector has declined in Rwanda, even though the absolute value in current terms has increased modestly. The share of government spending allocated to agriculture fell to less than 4 percent in recent years, compared with levels as high as 8.6 percent in 2002 (Table 6.1). Although the government's total spending grew more than 10 percent from 2001 to 2006, the growth rate of agricultural spending (in real terms) is negative for this period. The share of agricultural funding allocated to development is relatively high, averaging more than 12.3 percent

Table 6.1 Economic growth and government budget allocations, 1999-2006

Indicator	1999	2000	2001	2002	2003	2004[a]	2005[a]	2006	Growth rate (%)
Value (1999 constant billion RWF)									
AgGDP	270	283	295	330	333	339	359		4.2
Non-AgGDP	375	400	434	468	473	499	532		4.8
GDP	645	684	730	798	805	838	890		4.6
Agricultural spending			11.1	12.6	8.7	10.5	8.9	8.3	-6.5
Agricultural development spending	14.6	6.5	10.4	6.7	6.6	7.8	5.6	5.4	-5.8
Nonagricultural spending			168	134	215	249	250	244	11.8
Total spending	174	124	179	146	224	260	258	252	10.8
Total development spending	77	64	55	55	58	69	77	69	4.2
Ratio to GDP or total spending (%)									
Agricultural spending/ total spending			6.2	8.6	3.9	4.0	3.4	3.3	—
Agricultural spending/ AgGDP			3.8	3.8	2.6	3.1	2.5		—
Agricultural development spending/total development spending	19.0	10.2	18.9	12.3	11.4	11.4	7.2	7.8	—
Nonagricultural spending/ non-AgGDP			23.0	16.8	26.8	29.8	28.0		—
Total spending/GDP	27.0	18.1	24.5	18.3	27.8	31.0	29.0		—

Source: Authors' compilation using data from MINECOFIN (2006).
Notes: —, not applicable; AgGDP, agricultural gross domestic product; GDP, gross domestic product; RWF, Rwandan franc. Blank cells indicate data not available.
[a]Values are authors' estimates.

per year during 2000-06 versus 5.2 percent of agricultural spending in the national budget during this period. Nevertheless, even in this case, the share of resources allocated to agriculture has declined from an average of 16 percent in the early 2000s to less than 10 percent in recent years.

Estimate of Total Spending Required for Agricultural Growth

How much agricultural spending is really required to achieve CAADP and poverty reduction goals? We use a two-step approach to answer this question. The first step is to estimate the agricultural growth required to achieve development objectives using the so-called "poverty reduction elasticity." For example, to achieve the first MDG requires an annual growth rate of more than 6.5 percent in the agricultural sector between 2007 and 2020. The second step involves estimating the agricultural spending required to achieve the desired growth targets in agriculture. This relationship is termed the "elasticity of agricultural growth with respect to agricultural spending," and it can be estimated econometrically using historical data. The impact of many investments on growth cannot be realized immediately; hence, a comparatively long time series is needed to achieve a robust estimation. However, official national agricultural spending data are only available for 2001–06, so additional data (1995–2000) were drawn from the International Monetary Fund (IMF 2004).

The estimated elasticity of agricultural growth with respect to agricultural spending during 1995–2005 was 0.17—that is, for every 1 percent of growth in agricultural spending, 0.17 percent of growth in AgGDP results. This elasticity is much lower than the African average of 0.366, based on a cross-country estimation using a much longer data time series.[1] Because of Rwanda's recent history and the huge amount of spending required to re-establish basic conditions for agricultural production in areas affected by the genocide, the estimated coefficient between agricultural spending and agricultural growth may not represent the true relationship in the future. Moreover, many productive investment projects were initiated only recently, and their potential effects on agricultural growth cannot be captured in an econometric analysis.

For these two reasons, the elasticity based on the cross-country study is also used in calculating the required levels of public spending (Table 6.2). Two sets of values are reported, corresponding to two different agricultural growth scenarios: Scenario 21, corresponding to the CAADP target, and Scenario 22, corresponding to the first MDG (see Chapter 5 for further discussion of these scenarios). As discussed above, for 6.5 percent annual growth in AgGDP and a similar growth rate in the nonagricultural sector, total GDP will grow at 7.4 percent annually over the projection period in Scenario 22. The agricultural spending required under each scenario is reported in Table 6.2. A 6.3 percent increase in AgGDP per year from 2007 to 2015 requires associated growth in agricultural investment (table entry "agricultural development spending") of 35.9 percent annually (assuming low elasticity) or 18.4 percent (assuming high elasticity). Assuming that the government's allocation to nonagricultural sectors is proportional to nonagricultural GDP and that agricultural nondevelopment spending is proportional to AgGDP, the total government budget is estimated to grow 8.2 percent annually (assuming low elasticity) or 6.7 percent (assuming high elasticity).

Such estimations are consistent with the average growth in the total government budget in recent years, but agricultural spending requires much faster growth than in the past. Because of more rapid growth in agricultural

[1] The elasticity for the African average is from Fan, Yu, and Saurkar (2008).

Table 6.2 Estimated resource allocations to the agricultural sector: Model results, 2006-15 (percent)

		CAADP target		First Millennium Development Goal	
Indicator	Baseline (2001-06)	Low elasticity	High elasticity	Low elasticity	High elasticity
Growth rate					
AgGDP	4.2	6.2	6.2	8.8	8.8
Non-AgGDP	4.8	6.2	6.2	7.2	7.2
GDP	4.6	6.2	6.2	8.0	8.0
Agricultural spending	-6.5	30.3	15.2	45.6	22.6
Agricultural development spending	-5.8	35.9	18.4	52.3	26.8
Nonagricultural spending	11.8	6.3	6.3	7.4	7.4
Total spending	10.8	8.2	6.7	12.2	8.3
Agricultural spending/total spending					
2001-06 baseline value	4.92	—	—	—	—
For 2010	—	6.6	4.4	9.2	5.2
For 2015	—	17.6	9.5	34.5	12.0
Agricultural spending/AgGDP					
2001-06 baseline value	3.2	—	—	—	—
For 2010	—	4.7	3.0	6.3	3.5
For 2015	—	14.1	4.6	30.7	6.5
	2015	2015			
Nonagricultural spending/non-AgGDP	24.9	44.1	44.1	44.1	44.1
Total spending/GDP	26.1	32.1	28.3	38.3	27.9

Sources: Baseline values are authors' calculations based on data from MINECOFIN (2006); later values are model simulation results.

Notes: —, not applicable; AgGDP, agricultural gross domestic product; CAADP, Comprehensive Africa Agricultural Development Programme; GDP, gross domestic product.

spending than in total spending in these scenarios, the share of agricultural spending will rise to 4.4 (high elasticity) or 6.6 (low elasticity) percent in 2010 and 9.5 (high) or 17.6 (low) percent in 2015. Whether the government needs to meet the requirements of the Maputo declaration (African Union 2003) of allocating at least 10 percent of its total budget to agriculture depends on whether such spending can stimulate agricultural growth efficiently. With (less efficient) low elasticity, the government needs to allocate 18 percent of its total budget to agriculture by 2015, whereas if spending has (more efficient) high elasticity, about 10 percent of the total government budget would be needed to support 6 percent annual agricultural growth.

As mentioned previously, 6 percent annual agricultural growth for 9 years is insufficient for the country to meet the goal of halving national poverty by 2020; instead, a growth rate of 6.5 percent per year during 2007–20 is needed (the MDG scenario has a similar growth rate to that under Scenario 22 but uses the longer time period of 2007–20 rather than 2007–15). Estimates of the required spending to achieve this level of growth indicate that agricultural spending needs to grow at the extremely high rate of 45.6 percent annually between 2007 and 2015 for investment with low elasticity of growth or 22.6 percent for that with high elasticity (Table 6.2). Assuming the growth in nonagricultural spending is proportional to nonagricultural GDP and agricultural nondevelopment spending is proportional to AgGDP, total government spending would grow at 8.3 or 12.2 percent annually, depending on whether the elasticity was high or low. The share of agricultural spending would rise to 5.2–9.2 percent in 2010 and 12.0–34.5 percent in 2015, again based on elasticity. Although this rate of growth in public resources allocated to the agricultural sector seems unrealistically high, the resulting shares of agricultural spending are not uncommonly high based on experiences in many Asian countries in their early stages of development.

This analysis indicates one of the important challenges facing the Rwandan government in prioritizing public fund allocation. There is an inconsistency between the current budget allocation and the role of agriculture as the engine of overall growth and as the most important contributor to poverty reduction proposed in the government strategic documents. The good news is that the government has already recognized this inconsistency and has started to increase allocations to the agricultural sector and rural development in the second EDPRS for 2008–12 (MINECOFIN 2008). However, there is still a significant gap between the required increases and the planned ones in agricultural spending.

This analysis also raises an important issue of raising the efficiency of public spending. When spending efficiency is improved to that of the African average, growth in required agricultural spending as well as the share of agricultural spending in the total budget can be cut by more than half. To increase the efficiency of public investment requires further understanding of institutional and governance capacity and policy implementation process—important topics that are beyond the scope of the present study.

Assessing Investment Priorities by Estimating Potential Returns to Investment

Estimating the public resources needed to reach the overall agricultural growth target is important, but prioritizing investments is equally so. To prioritize the allocation of public investment, it is necessary to measure the returns to that investment. Public investment will generate externalities and directly and indirectly affect broad economic performance. Hence, the impact of public agricultural investment may be underestimated if the gains are narrowly measured for individual sectors that directly benefit through the interventions. Moreover, as we have shown in the previous chapter, agriculture-led growth is pro-poor, and the relationship between public investment in agriculture and poverty reduction needs to be taken into account when the impact of agricultural investment is assessed. In this section, we combine the public investment analysis with the DCGE model to assess the potential returns of agricultural investment from an economywide perspective.

As previously mentioned, because of the short period for available investment data in Rwanda, it is not possible to do any ex post analysis for evaluating the returns to public investment in agriculture. A different approach has to be applied so that the available data can be optimally used. With help from MINAGRI we have obtained a spending plan with detailed target and costing information, prepared for the second EDPRS (MINAGRI 2008). Based on this information and other data drawn from the literature, we have developed an ex ante approach that organizes the information as inputs into our DCGE model. We then use the model to conduct a series of simulation analyses to assess the returns to public investment in the agricultural sector.

Table 6.3 summarizes the total amount of public investment in agriculture planned for 2007–15, drawn from MINAGRI (2008). The original assessment of national needs to achieve the EDPRS targets is for a 5-year period (2008–12). To be consistent with the first MDG timeframe, we decided to consider a 9-year period (2007–15) for the analysis. Moreover, the initial assessment of the public investment required by the EDPRS targets had to be scaled down because of budget constraints set by the Ministry of Finance. The final agreed-on total for spending for agriculture is about 30 percent of what had been originally planned (Table 6.3).

However, for this analysis we decided to use the original budget allocation for three reasons. First, the purpose of this analysis is to provide an assessment of actual requirements in agricultural investment for achieving both the CAADP growth target and the first MDG of halving poverty by 2020. After assessing the growth requirement in Chapter 5, it is necessary to understand the cost of such growth without taking into account budget constraints. Second, our analysis considers a period of 9 years between 2007 and 2015. Without scaling down the total cost planned by MINAGRI, we can slightly reduce the annual cost by allocating the same amount of total spending in 9 years instead of 5 years (still, the annual spending is about 50 percent higher in our analysis than that included in the EDPRS; see the "Average annual total" column in Table 6.3 for comparison). Third, we also consider the CAADP targets, one of which requires African countries to allocate at least 10 percent of government spending to agriculture. The amount of agricultural spending,

Table 6.3 Total accumulated investments between 2007 and 2015: Model results

Agricultural spending used in the model (million RWF)	Cumulative total (2007-15)	Fixed investment	Recurrent investment	Fertilizer plus seed subsidy	Average annual total
Direct investment in crops	415,077	182,956	45,278	186,842	46,120
Forestry	8,641	6,742	1,899	–	960
Livestock	203,253	154,259	48,994	–	22,584
Research and development, extension	26,963		26,963	–	2,996
Rural finance	13,662		13,662		1,518
Other investments in horticulture	122,236	90,660	31,576		13,582
Other investments in traditional exports	22,840	16,940	5,900		2,538
Rural roads	21,965	21,965			2,441
Agricultural institutions	16,230		16,230		1,803
Other investments	11,265	11,263			1,252
Total	862,132	484,787	190,502	186,842	95,792
2007 annual total agricultural spending (million RWF)		13,517	6,434		19,951
Share in total government budget (%)	–	1.4	7.4		3.9
Proposed agricultural spending[a] (million RWF)	242,000	158,000	84,000		48,400
Share in total government budget (%)	7.0	11.5	4.1		
Increase from 2007 annual level (times)	–	–	–	–	1.4
Annual spending on agriculture[b] (million RWF)					69,143
Compared with the number used in model, without fertilizer subsidy[c]					1.09
Compared with the number used in model, with fertilizer subsidy[c]					1.39

Source: Authors' calculations based on data from MINAGRI (2008).

Notes: –, not applicable; RWF, Rwandan franc. Blank cells indicate data not available.

[a]As proposed in EDPRS (MINECOFIN 2008) for 2008-12.

[b]Assuming that 10 percent of the total annual budget is devoted to agriculture.

[c]Calculated as the model value divided by the annual spending on agriculture, assuming that 10 percent of the total annual budget is devoted to agriculture.

if it is based on the original MINAGRI numbers, will allow the agricultural sector to reach 10 percent of total government budget (Table 6.3).[2]

The key information about budget allocation and targets to be achieved through such public investment (such as how many hectares of irrigated land or how many metric tons of improved seeds will be developed) is available in the MINAGRI (2008) dataset. Subsidies for fertilizer and improved seeds are not considered in the government budget allocation but are added as part of public spending in the MINAGRI budget allocation. As in many other African countries, the past poor performance of the agricultural sector in Rwanda is partly a consequence of the very limited use of improved seeds and fertilizers. According to Kelly et al. (2001b), average consumption of fertilizer per hectare of cultivated land is less than 4 kilograms in Rwanda. This contrasts sharply with present-day Sub-Saharan Africa as a whole (ranging from 9 to 11 kilograms per hectare). The average level of use of improved seed is also low, only about 1.3 percent of total seed applied, below the average for Sub-Saharan Africa (Fowler et al. 2007). Because of increased oil prices, the cost of fertilizer is so high that the majority of small farmers in Africa cannot afford it. Thus a partial subsidy on imports of fertilizer is necessary to increase the use of fertilizer in crop production.

The above budget allocation information is important for an analysis assessing the economic outcome of public investment allocation; however, such information alone is not enough. To assess the impact of public investment on growth and poverty reduction ex ante, it is necessary to link the investment and its targets to the economic activities of the agricultural sector. As shown in Appendix A Table A.1, the agricultural sector is composed of 16 crop subsectors, 7 livestock subsectors, fishery, and forestry. It is a big challenge to allocate investment in specific agricultural subsectors (especially for the crop sectors) such that investment can be directly linked to the economic outcome in the ex ante analysis. To get such detailed information at the agricultural subsector and crop level is beyond the current planning capacity of MINAGRI, and we have to use other information to do it. We conducted a broad literature review to obtain such information as the current level of crop yields in the country, the unit production revenue per hectare of different crops, the achievable level of crop yields, and recommended levels of fertilizer and improved seed. These and other data are used in our analysis. Using such information we allocated investments in specific agricultural subsectors according to the function of the investment. The results of such

[2] Seven percent of the total planned public spending between 2008 and 2012 will be allocated to agriculture in EDPRS (Table 6.3).

allocations are reported in Table 6.4. Some spending cannot be identified by specific sector (such as spending on agricultural research and development, extension, rural finance, and agricultural institutions). We decided to allocate such spending according to the ratio of total nonspecific spending to the total for specific-sector investment (such as for terracing, irrigation, marshland development, and fertilizer and improved seed subsidies). The last row of Table 6.4 reports the targeted outcome of each type of investment in terms of hectares affected.

Growth in agricultural production at the crop or subsector level is unlikely to result from a single intervention—many interventions are needed for maximum impact. For example, without the application of fertilizer and improved seeds, returns to irrigation investment are low. To capture the joint impact of different types of investments, we need to further combine different types of public investment at the agricultural subsector level. To do so we first assume that once modern inputs are applied, the level of their applications will be consistent with the recommended level to achieve the maximum returns from such interventions. The recommended level of fertilizer is drawn from Fleskens (2007). We further assume that once the cropland is irrigated, farmers will first apply fertilizer and improved seeds on such land to get maximum returns from such combinations. With these assumptions, we can identify which intervention is a binding constraint for the combination of modern technologies (inputs). As shown in Table 6.4 (last row), the outcome of investment in irrigation will be 55,000 hectares of crop area, which is the least for the four types of interventions studied (terracing, irrigation, and application of fertilizer and improved seeds). Thus we decided to allocate other interventions following the allocation of irrigation area.[3]

The first columns of Tables 6.5 and 6.6 correspond to the allocation of irrigation area by crops in the base year (2006) and by 2015, respectively. All irrigated areas use fertilizer and improved seed, called "Modern" technology in the column heads. After irrigation, the second type of investment that may also result in the use of fertilizer and improved seed is terracing. We assume that farmers will also use fertilizer and improved seeds on their terraced lands. The second columns of Tables 6.5 and 6.6 report land allocations by crop to the combination of terracing, fertilizer, and improved seed; the increased total amount of terraced land is consistent with investment in terracing.

In the remaining columns of these tables, fertilizer seldom appears for most crops because of the constraints of planned government spending on fertilizer subsidies. That is, given the currently high prices for fertilizer and

[3] Marshland development is only for rice and some export crops (for example, sugarcane).

seeds, we assume that without certain government supports farmers will be unlikely to increase their use of such inputs to the recommended level.[4] Tables 6.7 and 6.8 report the share of such allocation reported in Tables 6.5 and 6.6, respectively.[5]

The second to last columns of Tables 6.5 and 6.6 report land allocations without applying any modern inputs. The ratio of traditional to total land by crops is also calculated. It is a well known fact that the use of modern inputs is still very low in Rwanda. More than 90 percent of its land is farmed using traditional farming technology in which no modern inputs are applied. However, the ratio of traditional to total area farmed varies across crops. In general, the high-value crops and rice use more modern inputs than do other crops. The last row of Table 6.8 reports the share of traditional to total land by 2015, which will decline if the planned public investment is implemented over the projection period. This ratio will fall to 76 percent by 2015, indicating the increasing use of modern inputs in crop production through public intervention. However, after 9 years of investment in irrigation, the irrigated land area will still be small, accounting for 4.4 percent of all croplands (Table 6.8). Only 17.6 percent of land will be improved through investment in radical terracing, whereas marshland development—mainly for growing rice and other export crops (such as sugarcane)—will amount to about 1.7 percent of cropped areas. Through subsidies, fertilizer (if applied to the recommended level) will be used on about 8 percent of land,[6] and the application of improved seeds will cover 14 percent of land. Note that actual land allocation in the model is endogenous, so that demand-side factors and changes in relative price will also affect allocation.

With the use of modern inputs, yields are expected to rise. Table 6.9 reports the projected yields resulting from public investment into different combinations of modern inputs. Our literature search was used to define standard yields for different types of modern inputs and their combinations. Moreover, the land allocated to such investment combinations reported in Tables 6.5–6.8 is consistent with the yields reported in Table 6.9. Such yield levels are often the result of applying modern inputs at their optimal levels.

---

[4] This assumption is consistent with the current fertilizer policy in Rwanda, in which about 50 percent of the price for imported fertilizer is covered by the government budget.

[5] According to Kelly et al. (2001b), 5 percent of farmers used inorganic fertilizers and/or lime on 3 percent of cultivated land during the first season of 2000. However, as we assume that the use of fertilizer has to be at the recommended level, the number reported in Table 6.7 for the fertilized land as share of cultivated land is smaller than in Kelly et al. (2001b).

[6] Sixteen percent of cultivated lands have the potential to be profitable (using the import prices for fertilizer from 2000) when the recommended doses of fertilizer are applied (Kelly et al. 2001b).

Table 6.4 Allocation of public investment by crops and livestock subsectors (million RWF)

Subsector	Terracing	Irrigation	Marshland development	Fertilizer subsidy, cumulative total (2007-15)	Fertilizer-related fixed investment	Seed subsidy, cumulative total (2007-15)	Seed-related fixed investment	Total
Grains								
Wheat	1,707	556	–	714	–	560	–	3,538
Maize	9,342	2,177	–	2,808	–	2,851	–	17,178
Paddy rice	–	1,439	50,172	5,375	–	12,773	–	69,759
Sorghum	9,031	369	–	1,759	–	2,326	–	13,486
Roots and tubers								
Irish potatoes	9,686	1,076	–	48,094	–	2,663	–	61,520
Sweetpotatoes	5,828	467	–	24,414	–	1,558	–	32,266
Cassava	6,368	483	–	25,743	–	1,695	–	34,289
Other roots	1,234	36	–	5,101	–	314	–	6,686
Other crops								
Pulses	11,117	157	–	1,292	–	2,790	–	15,356
Vegetables	–	3,698	–	14,174	45,330	23,352	45,330	131,883
Bananas	66,564	857	–	17,990	–	16,685	–	102,097
Fruits (excluding bananas)	–	1,434	–	2,532	–	355	–	4,320
Oilseeds	7,275	510	–	892	–	1,927	–	10,604

Export crops								
Coffee	–	11,650	–	22,847	5,795	5,751	5,795	51,837
Tea	–	6,408	–	11,565	2,675	2,910	2,675	26,233
Other export crops	–	228	15,858	1,542	–	4,243	–	21,872
Livestock								
Cattle	–	–	–	–	–	–	–	71,243
Sheep and goats	–	–	–	–	–	–	–	6,938
Swine	–	–	–	–	–	–	–	4,741
Poultry	–	–	–	–	–	–	–	2,419
Raw milk	–	–	–	–	–	–	–	133,232
Eggs	–	–	–	–	–	–	–	2,481
Other livestock	–	–	–	–	–	–	–	910
Fishing	–	–	–	–	–	–	–	4,560
Forestry	–	–	–	–	–	–	–	9,658
Total for crops	128,153	31,545	66,031	186,842	53,800	82,754	53,800	602,925
Total for livestock	–	–	–	–	–	–	–	221,964
Total sum								839,107
Targeted area by 2015 (hectare)	226,000	55,000	20,000	112,505	–	169,918	–	–

Source: Authors' calculations based on data from MINAGRI (2008).
Notes: –, not applicable; RWF, Rwandan franc.

Table 6.5 Land allocation by type of investment/spending, 2006

Subsector	Treatment (thousand hectares)										
	Modern (irrigation, fertilizer, and seed)	Terracing, fertilizer, and seed	Terracing and seed	Terracing	Marshland development, fertilizer, and seed	Marshland development and seed	Marshland development	Seed	Traditional	Total	Traditional/ total (%)
Grains											
Wheat	46	28	1,482	46	–	–	–	82	21,287	22,972	92.7
Maize	230	481	6,870	230	–	–	–	399	106,627	114,836	92.9
Paddy rice	2,105	–	–	–	559	3,181	2,575	–	5,613	14,033	40.0
Sorghum	170	256	10,816	170	–	–	–	592	158,294	170,298	93.0
Roots and tubers											
Irish potatoes	1,398	2,598	496	9,033	–	–	–	–	126,226	139,750	90.3
Sweetpotatoes	139	21	4,299	2,368	–	–	–	–	131,899	138,725	95.1
Cassava	119	137	3,564	3,939	–	–	–	–	111,101	118,860	93.5
Other roots	13	6	793	418	–	–	–	–	24,021	25,251	95.1
Other crops											
Pulses	388	635	3,570	2,684	–	–	–	–	380,244	387,521	98.1
Vegetables	4,262	–	–	–	–	–	–	–	46,430	50,692	91.6
Bananas	916	533	10,323	916	–	–	–	–	353,608	366,296	96.5
Fruits (excluding bananas)	1,918	–	–	–	–	–	–	–	32,218	34,138	94.4
Oilseeds	59	325	1,623	59	–	–	–	–	56,495	58,560	96.5
Export crops											
Coffee	8,523	–	–	–	–	–	–	–	20,239	28,762	70.4
Tea	4,688	–	–	–	–	–	–	–	8,592	13,280	64.7
Other export crops	653	–	–	–	114	1,534	3,142	–	2,717	8,160	33.3
Total	25,623	5,020	43,837	19,862	673	4,715	5,717	1,075	1,585,612	1,692,134	93.7

Source: Authors' calculations based on data from MINAGRI (2008).
Note: –, not applicable.

Table 6.6 Targeted land allocation by type of investment/spending, 2015

Subsector	Treatment (thousand hectares)									
	Modern (irrigation, fertilizer, and seed)	Terracing, fertilizer, and seed	Terracing and seed	Terracing	Marshland development, fertilizer, and seed	Marshland development and seed	Marshland development	Traditional	Total	Traditional/ total (%)
Grains										
Wheat	1,014	627	2,157	1,774	–	–	–	19,781	25,352	78.0
Maize	4,019	3,749	14,049	6,200	–	–	–	97,569	125,586	77.7
Paddy rice	4,610	–	–	–	7,056	9,231	5,224	4,610	30,731	15.0
Sorghum	813	1,221	20,560	5,332	–	–	–	134,652	162,578	82.8
Roots and tubers										
Irish potatoes	3,270	5,229	5,624	20,219	–	–	–	114,308	148,650	76.9
Sweetpotatoes	951	735	6,779	9,416	–	–	–	128,444	146,325	87.8
Cassava	959	1,781	5,447	11,602	–	–	–	107,581	127,370	84.5
Other roots	76	136	1,481	1,769	–	–	–	25,800	29,261	88.2
Other crops										
Pulses	661	2,609	5,539	18,278	–	–	–	385,815	412,901	93.4
Vegetables	10,697	–	–	–	–	–	–	42,322	53,019	79.8
Bananas	2,408	9,784	28,042	90,923	–	–	–	254,119	385,276	66.0
Fruits (excluding bananas)	4,413	–	–	–	–	–	–	31,293	35,705	87.6
Oilseeds	947	1,947	3,762	7,603	–	–	–	48,861	63,120	77.4
Export crops										
Coffee	28,800	19,793	–	–	–	–	–	–	48,593	0.0
Tea	15,840	6,596	–	–	–	–	–	–	22,436	0.0
Other export crops	1,049	–	–	–	2,033	4,368	3,193	287	10,929	2.6
Total	80,526	54,207	93,441	173,115	9,089	13,599	8,417	1,395,439	1,827,833	76.3

Source: Authors' calculations based on data from MINAGRI (2008).
Note: –, not applicable.

Table 6.7 Share of land allocation by type of investment/spending, 2006 (percent)

Subsector	Treatment								
	Modern (irrigation, fertilizer, and seed)	Terracing, fertilizer, and seed	Terracing and seed	Terracing	Marshland development, fertilizer, and seed	Marshland development and seed	Marshland development	Seed	Traditional
Grains									
Wheat	0.2	0.1	6.5	0.2	–	–	–	0.4	92.7
Maize	0.2	0.4	6.0	0.2	–	–	–	0.3	92.9
Paddy rice	15.0	–	–	–	4.0	22.7	18.4	–	40.0
Sorghum	0.1	0.2	6.4	0.1	–	–	–	0.3	93.0
Roots and tubers									
Irish potatoes	1.0	1.9	0.4	6.5	–	–	–	–	90.3
Sweetpotatoes	0.1	0.0	3.1	1.7	–	–	–	–	95.1
Cassava	0.1	0.1	3.0	3.3	–	–	–	–	93.5
Other roots	0.1	0.0	3.1	1.7	–	–	–	–	95.1
Other crops									
Pulses	0.1	0.2	0.9	0.7	–	–	–	–	98.1
Vegetables	8.4	–	–	–	–	–	–	–	91.6
Bananas	0.3	0.1	2.8	0.3	–	–	–	–	96.5
Fruits (excluding bananas)	5.6	–	–	–	–	–	–	–	94.4
Oilseeds	0.1	0.6	2.8	0.1	–	–	–	–	96.5
Export crops									
Coffee	29.6	–	–	–	–	–	–	–	70.4
Tea	35.3	–	–	–	–	–	–	–	64.7
Other export crops	8.0	–	–	–	1.4	18.8	38.5	–	33.3
Total	1.5	0.3	2.6	1.2	0.0	0.3	0.3	0.1	93.7

Source: Authors' calculations based on data from MINAGRI (2008).
Note: –, not applicable.

Table 6.8 Share of targeted land allocation by type of investment/spending, 2015 (percent)

	Treatment							
Subsector	Modern (irrigation, fertilizer, and seed)	Terracing, fertilizer, and seed	Terracing and seed	Terracing	Marshland development, fertilizer, and seed	Marshland development and seed	Marshland development	Traditional
Grains								
Wheat	4.0	2.5	8.5	7.0	–	–	–	78.0
Maize	3.2	3.0	11.2	4.9	–	–	–	77.7
Paddy rice	15.0	–	–	–	23.0	30.0	17.0	15.0
Sorghum	0.5	0.8	12.6	3.3	–	–	–	82.8
Roots and tubers								
Irish potatoes	2.2	3.5	3.8	13.6	–	–	–	76.9
Sweetpotatoes	0.7	0.5	4.6	6.4	–	–	–	87.8
Cassava	0.8	1.4	4.3	9.1	–	–	–	84.5
Other roots	0.3	0.5	5.1	6.0	–	–	–	88.2
Other crops								
Pulses	0.2	0.6	1.3	4.4	–	–	–	93.4
Vegetables	20.2	–	–	–	–	–	–	79.8
Bananas	0.6	2.5	7.3	23.6	–	–	–	66.0
Fruits (excluding bananas)	12.4	–	–	–	–	–	–	87.6
Oilseeds	1.5	3.1	6.0	12.0	–	–	–	77.4
Export crops								
Coffee	59.3	40.7	–	–	–	–	–	–
Tea	70.6	29.4	–	–	–	–	–	–
Other export crops	9.6	–	–	–	18.6	40.0	29.2	2.6
Total	4.4	3.0	5.1	9.5	0.5	0.7	0.5	76.3

Source: Authors' calculations based on data from MINAGRI (2008).
Note: –, not applicable.

Table 6.9 Yield by type of investment/spending, 2006–15 (metric tons/hectare)

	Treatment										
Subsector	Modern (irrigation, fertilizer, and seed)	Terracing, fertilizer, and seed	Terracing and seed	Terracing	Marshland development, fertilizer, and seed	Marshland development and seed	Marshland development	Traditional	Baseline yield, 2006[a]	Average yield in baseline run, 2015[a]	Targeted average yield, 2015[a]
Grains											
Wheat	3.0	2.8	1.4	2.3	–	–	–	1.2	0.8	1.2	1.6
Maize	3.5	3.1	1.5	2.5	–	–	–	1.2	0.8	1.0	1.5
Paddy rice	6.4	5.8	5.5	5.8	6.1	–	6.1	5.5	3.7	4.9	4.7
Sorghum	3.4	2.7	1.7	2.4	–	–	–	1.5	1.1	1.5	1.7
Roots and tubers											
Irish potatoes	27.2	24.9	15.5	20.2	–	–	–	12.6	9.1	11.9	14.7
Sweetpotatoes	15.0	16.8	11.2	11.5	–	–	–	7.7	5.5	6.9	7.8
Cassava	27.3	24.7	11.2	18.8	–	–	–	8.6	6.2	8.0	10.5
Other roots	24.5	17.3	9.6	12.3	–	–	–	6.8	4.9	6.0	6.8
Other crops											
Pulses	2.7	2.4	1.2	2.1	–	–	–	1.1	0.8	1.0	1.0
Vegetables	29.5	19.6	14.0	19.6	–	–	–	14.0	9.4	7.3	9.4
Bananas	17.9	15.7	10.5	14.9	–	–	–	9.9	7.2	10.0	12.9
Fruits (excluding bananas)	28.6	19.1	13.6	19.1	–	–	–	13.6	9.7	14.3	17.1
Oilseeds	2.7	1.8	1.1	1.4	–	–	–	0.8	0.6	0.8	1.1
Export crops											
Coffee	1.0	0.8	0.7	0.8	–	–	–	0.7	0.7	1.1	1.1
Tea	1.7	1.4	1.2	1.4	–	–	–	1.2	1.1	1.7	1.7
Other export crops	45.8	38.2	27.3	38.2	32.7	23.8	32.7	27.3	17.6	26.5	35.0

Sources: Authors' calculations based on data from Clay et al. (1995); Clay, Reardon, and Kangasniemi (1998); Kelly et al. (2001a, 2001b); MINAGRI (2004a); ICARRD (2006); Fleskens (2007); and Alliance of CGIAR Centers (2008).

Note: –, not applicable.

[a]Yield is weighted by area.

The yields using modern technologies can be quite high; for example, maize can reach 3.5 and Irish potatoes 27.2 metric tons per hectare.

Although these yields are high, when weighted by their corresponding land allocations, their effects on national average yields are small: only about 4.7 percent of total cropland is expected to achieve such high yields (Table 6.8). Moreover, more than 70 percent of cropland will still be without access to modern inputs by 2015; thus traditional technology will continue to play a dominant role in determining national average yields for most crops. As shown in the last column of Table 6.9, yields from most crops will continue to be low by 2015, even after significantly increased public investment in agriculture. For example, the average maize yield for the country is expected to be only about 1.5 metric tons per hectare by 2015, although it is possible to have a yield of 3.5 metric tons per hectare by using irrigation, fertilizer, and improved seeds. Although the projected national average is a significant increase from the current maize yield (0.8 metric tons per hectare), it is lower than what was already achieved by many Sub-Saharan African countries. To compare the baseline and projected yields (by 2015) assuming business as usual, three levels of national average yields are shown in Table 6.9. The yields reported in the "Baseline yield, 2006" column are consistent with the baseline-year (2006) yields applied in the model. Values given in the "Average yield in baseline run, 2015" column are consistent with those in Table 4.1, and the numbers reported in the "Targeted average yield, 2015" column can be understood as the projected results assuming the implementation of all planned public investments.

Targeted crop average yields and the outcome of public investment in livestock and roads are applied in the DCGE model as exogenous shocks to the land productivity (in the case of crop production) or total factor productivity (in the case of livestock production, trade, transportation, and communication sectors). Ideally, the DCGE model should consider all these technologies to endogenously capture the national productivity growth that results from enhanced yields. However, because of a lack of disaggregated information at the district and farm levels (the production functions are defined at the district level for small and medium-large farm groups in the model), we are unable to capture the endogenous productivity growth in the analysis. Instead, we have to apply the exogenous productivity shocks calculated from the potential growth in the yields as results of public investments and use of modern inputs. With such exogenous shocks, the DCGE model generates results that are in general consistent with the planned public investment and its targets. Thus we do not need to report again the model results for crop yields.

Table 6.10 reports model results for land allocation under the investment and baseline scenarios. Although total cultivated land increases by 1.9 percent by 2015 in the investment simulation over that in the baseline run for same year, land areas are reduced for some foodcrops because of improvements in their yields as a result of investment. For example, sorghum is currently one of the most widely grown staple crops (170 thousand hectares). In the investment scenario, its cultivated area falls to 164 thousand hectares by 2015. Similarly, the area harvested for sweetpotatoes falls to 123 thousand hectares in the investment scenario from its current area of 139 thousand hectares. Such declines are the endogenous results of the model. The population grows at 2.5 percent annually in this period in the model simulation. Increased demand stemming from population growth and additional per capita consumption at higher income levels is primarily met by improvements in land productivity instead of the expansion of cultivated land. With little additional land brought into cultivation for feeding more people, this result indicates the importance of land productivity in the case of staple crop production for environmentally sustainable agricultural growth.

However, land areas devoted to many high-value and export crops increase by 16 percent in the same period. All export crops together account for less than 10 percent of total cultivated land by 2015, but they account for more than 25 percent of land expansion. Considering the environmental sensitivity of land expansion in the long-term growth of Rwandan agriculture, a balance between promoting growth in export-oriented agriculture and controlling area expansion to reduce land degradation will be a policy challenge for the government.

The land allocation results (Table 6.10) together with those for increases in yields (Table 6.9) can be used to determine the subsector levels of agricultural growth (Table 6.11) and the growth in the economy as a whole and in aggregated economic sectors (Table 6.12). The growth results under the investment scenario (Table 6.11) are comparable with those in Tables 5.2 and 5.3 for 2015 under the agricultural growth scenario (Scenario 21). Similar growth rates in land productivity are obtained under the two scenarios: one considering growth targets only and the other using investments to support such targets.

Moreover, increased public investment, if implemented as planned, is able to support more than 6 percent annual growth in AgGDP over 2007–15 (Table 6.12), which is similar to the growth rate obtained under the agricultural growth scenario (Scenario 21). Livestock will have the highest growth rate, more than doubling its 2007 annual growth rate. Among crops, cash and export crops benefit most from increased investment (7.9 percent annual growth), followed by the grain sector (7.0 percent). Even though agricultural investment is not applied directly to the nonagricultural sectors, the food-processing sectors will benefit indirectly from such investment. Conversely, investment in roads

Table 6.10 Land allocation: Model results under the investment scenario (thousand hectares)

Crop	2006 level	2015 Baseline run	2015 Investment scenario run
Grains			
Wheat	23	25	25
Maize	115	126	126
Paddy rice	14	18	30
Sorghum	170	162	164
Roots and tubers			
Irish potatoes	140	149	154
Sweetpotatoes	139	136	123
Cassava	119	123	111
Other roots	25	29	30
Other crops			
Pulses	388	410	428
Vegetables	51	51	51
Bananas	366	393	394
Fruits (excluding bananas)	34	32	33
Oilseeds	59	61	60
Export crops			
Coffee	29	34	45
Tea	13	16	24
Other export crops	8	9	12
Total	1,692	1,775	1,809

Sources: Numbers for 2006 are authors' calculations based on data from MINAGRI (2007); numbers for 2015 are model simulation results under the baseline and investment scenarios.

will improve productivity in the trade, transport, and communication sectors, which in turn helps growth in the agricultural sector.

Table 6.13 shows the impact of public investment on poverty rates. Through its linkages with growth, poverty falls more in the rural areas than in urban ones. As discussed in Table 5.5, the calculated poverty-growth elasticity shows that staple-led agricultural growth is more pro-poor than other agricultural growth: 1 percent of additional growth in the economy led by growth in staple-crop productivity results in 1.4-2.4 percent poverty reduction. Although the 36.2 percent poverty rate in 2015 is still higher than the first MDG, the difference is not unbridgeable.[7] As mentioned in the previous chapter, with additional growth from the nonagricultural sector, the country will be able to achieve the first MDG by 2020.

[7] The national poverty rate was 60.3 percent in 1999 (MINECOFIN 2003). Thus to reach the first MDG target, the poverty rate must fall to 30 percent by 2015.

Table 6.11 Annual growth rates for agricultural subsectors: Model results under the investment scenario (percent)

Subsector	Baseline run (2006)	Investment scenario run (2015)
Grains		
Wheat	4.9	8.5
Maize	3.9	8.2
Paddy rice	3.9	9.7
Sorghum	3.1	4.2
Roots and tubers		
Irish potatoes	3.6	6.5
Sweetpotatoes	2.1	2.4
Cassava	3.2	5.2
Other roots	3.9	5.7
Other crops		
Pulses	3.4	4.6
Vegetables	4.4	7.5
Bananas	3.7	6.6
Fruits (excluding bananas)	3.5	5.4
Oilseeds	3.7	6.6
Export crops		
Coffee	6.3	9.8
Tea	6.1	10.5
Other export crops	4.5	10.1
Livestock		
Cattle	4.2	7.4
Sheep and goats	4.5	9.6
Swine	3.8	7.5
Poultry	4.6	11.8
Raw milk	7.3	17.8
Eggs	7.5	18.0
Other livestock	7.0	18.8
Forestry	3.8	6.5
Fishing	4.8	8.9

Source: Numbers are model simulation results under the baseline and investment scenarios.

We also calculated the dollar-to-dollar returns to public investment in agriculture. Such returns are measured by increased GDP and AgGDP, both in real terms (in base-year prices). To do so, we must also consider future returns after 2015 from additional investment occurring over the projection period 2007–15. To calculate such future returns, we applied a 10 percent social discount rate and chose a very long time period (such as 20–30 years), such that for a given amount of public investment (together with recurrent spending), the returns are eventually discounted to close to zero. We included spending related to fertilizer and seed subsidies in the recurrent spending, which is also discounted by 10 percent for the same period. As mentioned

Table 6.12 Annual growth rates in aggregated economic sectors and GDP: Model results under the investment scenario (percent)

Sector/indicator	Baseline run (2006)	Investment scenario run (2015)
GDP	4.8	6.0
AgGDP	3.8	6.5
Grains	3.6	7.0
Root crops	3.1	5.1
Other foodcrops	4.1	6.5
Cash and exportable crops	4.7	7.9
Livestock	5.5	12.6
Fishing and forestry	4.0	6.9
Industry	5.1	5.1
Processing and manufacturing	6.2	7.9
Coffee processing	8.3	11.8
Tea processing	6.7	10.6
Services	5.7	5.8
Trade	4.4	6.2
Transport	7.0	6.5
Communications	8.0	6.4

Source: Numbers are model simulation results under the baseline and investment scenarios.

Notes: AgGDP, agricultural gross domestic product; GDP, gross domestic product.

Table 6.13 Poverty rates: Model results under the investment scenario (percent)

Region	2006 level	2015	
		Baseline run	Investment scenario run
National	57.6	46.7	37.2
Urban	28.8	27.9	19.9
Rural	63.4	50.4	40.7

Sources: Numbers for 2006 are authors' calculations based on data from MINAGRI (2007); numbers for 2015 are model simulation results under the baseline and investment scenarios.

earlier, returns to public funds invested in a specific sector should not be measured solely by the benefits to this sector. Public investments often generate strong and positive externalities in the economy, such as technological spillovers in agriculture and growth linkages between agricultural and nonagricultural activities. Thus increases in GDP over time (discounted to the current value) are used to measure the returns to agricultural investment (Table 6.14), allowing us to fully capture the economywide gains of public investment in agriculture.

It is also important to measure the economywide returns as results of public funds invested in a specific agricultural subsector to determine which sector is most efficient in its use of investment. For this purpose, we designed another 21 model scenarios, each focused on the investment going to a specific sector. For example, Table 6.14 shows the result of public funds invested in maize production ("Maize" entry). In this scenario, we assume that only productivity growth in maize production is the direct outcome of increased investment, holding productivity growth in all other sectors at their baseline-run levels.

Table 6.14 Returns to public investment by agricultural subsector: Model results under the investment scenario, 2006–15

Subsector	GDP/investment	AgGDP/investment
Grains		
Maize	7.02	6.59
Paddy rice	1.41	1.22
Wheat	5.34	5.15
Roots and tubers		
Cassava	5.46	4.61
Irish potatoes	5.88	5.66
Sweetpotatoes	2.53	2.22
Other crops		
Pulses	9.09	8.21
Bananas	5.35	4.94
Oilseeds	5.89	4.73
Export crops		
Coffee	1.01	1.74
Tea	1.95	2.52
Other export crops	1.08	1.07
Livestock		
Poultry	10.54	10.09
Other livestock	1.81	1.74
Fishing	12.50	12.35
Grains	2.75	2.73
Root crops	5.03	4.65
Cash and export crops	1.02	1.24
Livestock	2.02	1.90
Staple crops and livestock	3.84	3.63
Agriculture, total	3.19	3.11

Source: Numbers are model simulation results derived from authors' calculations using 2006 data from MINAGRI (2007).

Notes: AgGDP, agricultural gross domestic product; GDP, gross domestic product.

We also consider productivity growth for groups of commodities in Table 6.14, specifically grains, root crops, cash and export crops, and livestock. For example, the entry for "Grains" includes yield increases coming directly from public investment in grain production (including maize, rice, wheat, and sorghum production) while investment in other sectors is held at baseline-run levels. In "Staple crops and livestock," productivity growth in these subsectors results from public investment in them while holding productivity growth in export crops at the baseline-run level; in "Agriculture, total" agricultural investment in all subsectors is considered.

The "GDP/investment" column of Table 6.14 shows the increment in GDP for each unit of increased public funding invested in the agricultural sector or subsector. These returns are equivalent to benefit/cost ratios. The "AgGDP/investment" column lists the ratio of incremental AgGDP to that of investment. As shown in the final entries of the "AgGDP/investment" column, economywide returns to public investment in overall agriculture are 3.11:1 and are 3.63:1 from investment in staple production (including both staple crops and livestock). We also notice that the returns to the investment measured as increased GDP are higher than those measured as increased AgGDP for the same amount of public spending in agriculture. These results further indicate the importance of measuring the complete economywide returns to agricultural investment, which also indirectly benefits the nonagricultural sectors through production and consumption linkages, as discussed earlier in this report.

At the aggregated subsector level (grains, root crops, cash and export crops, and livestock), returns to investment in root crops are highest, measured either as increased GDP or AgGDP. Returns to investment in grain production rank second, and returns to investment in export crops are the lowest, but they are still greater than one. For all agricultural sectors ("Maize" through "Fishing"), returns to investment, as measured by increased GDP, are greater than the cost of the investment: all returns are greater than one and many are greater than two or three.

The returns to investment for coffee (1.01) are the lowest for all commodities listed in the table. However, measured directly by increased AgGDP, coffee investment is quite profitable, as an investment of US$1.00 results in an increment of US$1.74 in AgGDP over time. That the increased GDP is significantly less than the increased AgGDP indicates the weak linkage of the coffee sector with the rest of the economy. Coffee is mainly grown for export, and coffee processing is predominantly simple. Without increasing the value added of coffee production, the sector is unlikely to create strong linkages and multiplier effects in the economy. Moreover, export-led growth can have certain macroeconomic effects on the real exchange rate. For example, if growth in exports of primary products (including primary processed agricultural products, such as coffee and tea) is too high, it may create a "Dutch disease" effect that raises the real exchange rate in the country. An appreciated exchange rate could in turn raise the cost for other exportable sectors and make them more difficult to export. An appreciated exchange rate causes imports to become cheaper and thus import-competitive sectors become unable to compete. Thus the weak linkage of coffee to the rest of the economy and this "Dutch disease" effect are important factors contributing to the low returns to investment in coffee as measured by increases in GDP.

CHAPTER 7

Discussion and Conclusions

Understanding alternative options for agricultural growth and their linkages with poverty reduction, and prioritizing agricultural investments are the two key components of any agricultural development strategy. However, the relationships between growth and poverty reduction and between targeted growth and required public investment are not straightforward—solid research is needed to support evidence-based policymaking and strategy formulation. This report provides such a study for Rwanda. An economywide DCGE model was developed and applied to the most recent economic data and public investment information to analyze the agricultural growth and investment options for poverty reduction in Rwanda. The report focuses on analyzing the linkages and trade-offs between growth and poverty reduction goals at the macro-, meso- and microeconomic levels and helps to provide answers to three policy-related questions:

1. What are the most effective pro-poor options for agricultural growth?
2. How can important linkages between agricultural and nonagricultural sectors be strengthened such that nonfarm activities will become an important source of income to encourage both growth and poverty reduction?
3. What are the most cost-effective public investment choices for stimulating shared growth and poverty reduction?

In trying to provide answers to these three questions, the analysis of this report focuses mainly on the "what" issues. Such questions are important for evidence-based agricultural strategizing, but they are not enough to identify binding constraints in development processes and hence to guide the strategy to be better targeted and more practical. For example, it is unlikely for us to quantitatively evaluate the returns to past investments in Rwanda because of the country's history (including many interruptions in its development process by civil conflicts and the 1994 genocide). Our analysis is thus unable to measure and rank the constraints to agricultural development (which are often manifold, given the country's history) and to prioritize public investments accordingly. In addition we have not addressed the pressing need to assess the challenges of political economy faced by the government and its institutional and governance capacities to effectively implement any strategy. In measuring the contributions of specific subsectors to overall economic growth and poverty reduction, we are unable to take into account the environmental impacts of alternative options for agricultural growth and hence cannot prioritize public investment interventions to develop a sustainable strategy for agricultural development. Although these questions are also important for research on development strategies, they are beyond the scope of this report. Keeping these limitations of the report in mind, the following key messages can be drawn from its analysis.

The CAADP Goal of 6 Percent Annual Growth Is Not Only Necessary but Also Achievable

It is well known by researchers and policymakers in Rwanda and elsewhere that low agricultural productivity and its interaction with environmental sustainability are the most severe challenges to Rwandan agriculture and rural development. As Rwanda is a landlocked country with a low level of per capita income, pressure on land and hence on the environment is primarily due to the demands on food security stemming from rapid population growth. Even with 90–95 percent of cultivated land allocated to food production for domestic consumption, food supply is still tight for most staple crops, and some consumers have to depend on imports to meet their basic needs. Thus productivity-led growth in staple agriculture is a key component in the country's development strategy, and a series of targets have been planned in that strategy. Our DCGE model simulations indicate that the country's targeted agricultural subsector growth has to be met by increases in yield instead of land expansion. If such targets were achieved, it would allow Rwanda to meet the CAADP target of 6 percent AgGDP growth from 2008 to 2015 with little land expansion. Along with comparable growth in the nonagricultural sector, the agricultural growth rate would increase to 6.5 percent, and total GDP growth to 7.4 percent, as a result of economywide interlinkages. The model also analyzes the linkages of growth with the strategy's poverty reduction target, which shows that such growth would cause the national poverty rate to fall to 35.5 percent by 2015, a reduction of 24.8 percentage points over the 1999 rate. If this level of growth continues until 2020, Rwanda would be able to achieve the first MDG of halving its national poverty rate by 2020, a target included in the country's *Vision 2020* and EDPRS (MINECOFIN 2000, 2008).

The Pattern of Subsector Growth Matters

Agriculture is composed of different activities, and export subsectors often attract more attention in the strategic plans for agriculture of many African countries. On the strength of the belief that export-oriented subsectors can contribute more to agricultural transformation than can other agricultural subsectors, the former have also attracted disproportionately more public resources and favorable policies. A similar policy bias also exists in Rwanda. Thus it is necessary to understand the role of different agricultural subsectors in both growth and poverty reduction to help prioritize public investment and agricultural policy. Our DCGE model covers detailed agricultural activities (specified as individual subsectors) as well as a broad range of nonagricultural activities. Through linkages between agriculture and nonagriculture and among agricultural subsectors both in production and consumption processes, the DCGE model simulations show that most rural households benefit from rapid but broad-based agricultural growth, and the distribution of such benefits is comparatively equal. However, the most vulnerable households—those with extremely small landholdings and with few opportunities to participate in the production of cash crops—appear to benefit the least.

Analyzing the linkages between subsector-level growth and poverty reduction is one of the important contributions of this report, and such analyses were carried out by endogenously measuring the poverty reduction response to economywide growth led by different sectors. The report first compares the effect on poverty reduction of growth led by the agricultural sector versus that led by the nonagricultural sector. The model results show that growth in per capita GDP that is driven by agriculture has a much greater effect on poverty reduction than the same level of growth driven by the nonagricultural sector. The report further compares the subsectors' role in poverty reduction by simulating the economywide growth led by different agricultural subsectors. The simulation results indicate that overall growth driven mainly by increased productivity in staple crops has the greatest effect on reducing poverty. Agricultural households with better opportunities to produce high-value export products are better positioned to benefit from export agriculture. But these households are usually not as poor as other more remote households, so export-led growth may have less impact in reducing poverty.

The roles of different agricultural subsectors in economic development are also analyzed in terms of poverty reduction and food security. Cereals, especially rice and maize, are among the high priorities in the government's strategy document both for food security and for import substitution; accordingly, they have been assigned very high growth targets. If these growth targets are reached, cereals would become the most important source of income growth for many rural households, especially for those with the smallest landholdings. Our results also show that food security does not equal food self-sufficiency. Although growth in cereals would help the country reduce its dependence on imports, both maize and rice will continue to be imported, but the ratio of imports to domestic consumption will significantly fall by 2015.

Export agriculture is important in terms of its contribution to macro-economic balance in trade. High growth in both traditional and nontraditional agricultural exports is targeted in the government's strategy, and our model shows that such growth would significantly increase agricultural trade surpluses. The projected agricultural trade surplus would be about 29 percent of the nonagricultural trade deficit, more than three times the baseline ratio. This report does not explicitly measure the environmental impact of agricultural growth, but it does show that growth in export-oriented agriculture is often an outcome of land expansion. Although such crops currently account for less than 10 percent of total cultivated land, they can account for more than 25 percent of the increase in cultivated land over the next 10 years if public investment favors the development of such sectors. The balance between supporting export agriculture and controlling rapid land expansion in this sector will become increasingly important for the Rwandan government in its development of a strategy for environmentally sustainable agricultural growth.

The study also warns of possible price declines in some commodities with very high growth targets. Unbalanced growth does not always benefit producers if it is concentrated in a few subsectors. As the targeted growth rates for some livestock products (such as poultry and raw milk) are very high, a negative price effect could result if production growth is out of balance with income growth. Simulations indicate that with an annual growth in poultry production of more than 9 percent and in raw milk of more than 15 percent during 2007-15, the prices for these two commodities will decline 35-40 percent, even though their demand is highly income elastic.

Agricultural Spending Needs to Increase Substantially

In this report we calculated the public investment in agriculture that is required to meet the CAADP 6 percent agricultural growth goal and/or the poverty MDG by 2020. The efficiency of spending (which is captured in the elasticity of agricultural growth) affects the amount of agricultural spending needed to meet these goals. The report shows that meeting the CAADP growth target will require that the allocation of public resources to the agricultural sector rise to 10.0-17.6 percent of total spending by 2015, where the lower value reflects a higher elasticity of growth (more efficiency in spending) and the higher value reflects the lower elasticity. This level of allocation translates, in real terms, to 15-30 percent annual growth in agricultural spending over 2007-15. Given that agriculture needs to grow at 6.5 percent to meet the goal of halving the national poverty rate by 2020, more rapid growth in agricultural spending between 2008 and 2015 is required. Our result in terms of required agricultural spending (using the high elasticity that represents an average value for the Sub-Saharan Africa) is consistent with the CAADP target of a 10 percent minimum of government budget for agriculture. This result is also consistent with the original cost estimation prepared by MINAGRI (2007) without scaling down to meet the constraints set by the Ministry of Finance. Such a spending pattern seems to be necessary both for achieving more than 6 percent agricultural annual growth and for helping the country meet the target set in *Vision 2020* for halving poverty by 2020 (MINECOFIN 2000). Efficiency in public spending must also improve.

Economywide Returns to Public Investment in Agriculture Are High

For an agricultural strategy that can actually be implemented, it is necessary to understand what can be done in addition to what should be done. To help determine what is feasible, returns to public investment in agriculture need to be analyzed. Most cost-benefit analyses are at the micro or project level, which may significantly underestimate the returns to public investment: such investments generally have impacts going beyond the targeted sector or project. One of this report's important contributions is to measure such returns in an economywide perspective. Although the approach used is ex ante because of a paucity of past investment data needed for a sound econometric estimation, the results are informative for prioritizing public investment. We find that the economywide return to agricultural investment (or benefit/cost ratio) is 3.2, measured as increased GDP over time. Part of such gains comes from growth in the nonagricultural sector as a result of multiplier and linkage effects between agriculture and nonagriculture. Comparing the changes in GDP and AgGDP as the outcomes of US$1.00 of additional public investment in agricultural staples, we find that the increase in GDP is US$0.21 higher than that in AgGDP, indicating that returns to agricultural investment stimulate growth in the nonagricultural sector.

Highest Economywide Returns Come from Investment in Staple Crops

Using an approach similar to the one discussed above, we also measured returns to agricultural investments in different agricultural subsectors. The results show that economywide returns to public investment in staple crop production are the highest, and returns to root crop-related spending are the highest of all staple crops. Given that current yields and the use of modern inputs are extremely low for most grain and root crops (except for rice) in Rwanda, even compared to other Sub-Saharan African countries, the country needs to give a much higher priority to these crops than it has in the past. The high returns are also the result of a linkage effect, particularly for root crops: productivity growth in such crops lowers pressure on land expansion, which is an extremely important factor that must be taken into consideration in any agricultural strategy for Rwanda. Staple crops are broadly consumed by the poor in both rural and urban areas, and reduced food prices from productivity-led growth benefit poor consumers and create growth opportunities in other agricultural and nonagricultural sectors. However, the planned amount of investment in the Rwandan strategy document (MINECOFIN 2008) will not be enough to significantly improve the low yields for many crops nationwide over 2007–15. The average yield for maize may be as low as 1.5 metric tons per hectare by 2015, a level many African countries have already reached.

Investment in Export Crops: Trade-Offs between Exports and Overall Growth

Investment in export crops has the highest impact on growth in these subsectors, but the economywide returns are quite low. The growth rate of coffee and tea can reach double digits with targeted investment to the export sector as measured by increased GDP over time; however, such investment has the lowest returns of all agricultural spending, particularly in the case of coffee. The reason for these low returns is mainly due to the weak linkages with other economic activities on both the production and consumption sides. Although it is possible to strengthen such linkages in the export sector by promoting agroprocessing with high-value additions, because of weak income-consumption linkages, public investment in the export sector is not likely to become a dominant driver in overall economic growth or poverty reduction.

The export sector often attracts much government attention in many African countries. Government spending to promote the export sector represents a high share of total agricultural spending in relation to the size of the export sector with respect to the total agricultural economy. The findings of this report show relatively low economywide returns to public spending in the export sector, and relatively less poverty reduction from growth led by promoting exports, further indicating the importance of broad-based agricultural growth. Taking into account the linkages between agricultural growth and environmental sustainability (which have not been analyzed explicitly in this report), the role of the agricultural export sector, particularly that of large-scale, commercial agriculture (typical for tea production in Rwanda), needs to be rethought. Its growth is less constrained by domestic demand, and it is often land intensive. Agricultural development strategy, including an effective public investment strategy, should focus more on growth in which most farmers can participate and that is sustainable without further increasing land degradation and other environmental problems. Only such a strategy can be expected to be efficient and effective in growth and poverty reduction, as well as in the economic development of the country in general.

How Relevant Is This Report to Other African Countries?

Although Rwanda is one of the smallest countries and has the highest population density in Africa, similarities do exist between Rwanda and many other African countries. First, it is true that at the country level, few African countries have such high population pressures as in Rwanda. Constrained by little additional new land to support the country's high population growth, to develop an efficient agricultural system that is driven by growth in productivity and environmental sustainability is more urgent in Rwanda than elsewhere in Africa. However, recent rapid population growth in most other African countries has resulted in similar pressures on their agricultural development at the country or subnational level. Such countries as Ethiopia, Kenya, and Malawi face similar pressures in their most densely populated areas, which are often their "food baskets." Environment degradation, such as soil erosion and deforestation due to extensive farming systems, is commonly observed in some areas of almost all African countries, even in the countries with relatively abundant land, such as Ghana and Nigeria. Thus challenges facing Rwanda's agriculture today are becoming challenges to agricultural development in many other African countries. From this point of view, the analysis in this report in highly relevant to many African countries.

Second, with agriculture once again at the top of the agenda in economic development strategy for all African countries, an understanding of

1. the linkages in growth among agricultural subsectors and between agricultural and nonagricultural sectors and
2. the linkages between growth and poverty reduction

is commonly required for designing an agricultural strategy. Such understanding has to be country specific, but the methodology developed in this report, which takes an economywide perspective, is useful for many countries. The general principles that this report emphasizes are all relevant to other countries in designing their development strategy. These principles include the need to link agricultural subsector targets to overall growth and poverty reduction, to understand the various roles of different subsectors in growth and poverty reduction at the macro and micro levels, and to pay attention to certain trade-offs and negative effects (such as price effects) of rapid and unbalanced growth.

Third, almost all African countries are facing constraints on public resources for supporting agricultural growth; a financial gap between what should be done and what can been done is commonly seen in national development strategy documents. By calculating the public spending required for achieving the targets set by the strategy documents, this report explicitly measures such gaps and points out that bridging the financial gap has to be a part of any development strategy. The report emphasizes the relationship between spending efficiency and the amount of spending required to support agriculture, which is a common challenge facing many African countries. As in Rwanda, many African countries lack sufficient information to do a reasonable econometric analysis, so that the returns to public investment are unlikely to be measured at the sector level to better prioritize investment. This report provides a new approach (albeit one that is ex ante), which is able to fully utilize the information that is available in a country's PRSP preparation process. It is possible to adopt this approach in the strategic studies of other countries.

Finally, development strategy research at the country level in Africa is an emerging field in today's new political environment. As more and more countries have begun to develop their agricultural development strategies—and almost all such strategies are dynamic documents to be improved and updated over time—the needs for evidence-based development strategy research are high and demand is rising. However, researchers in this field are facing challenges stemming from the lack of solid data and the urgency of providing research outcomes in a relatively short time to meet policy demand. This report is far from perfect, but it is an encouraging attempt to make development research more relevant to the policies and strategies of developing countries.

APPENDIX A

Data on the Rwandan Economy and Household Demand Structure

Figure A.1 GDP and AgGDP, 1981-2006

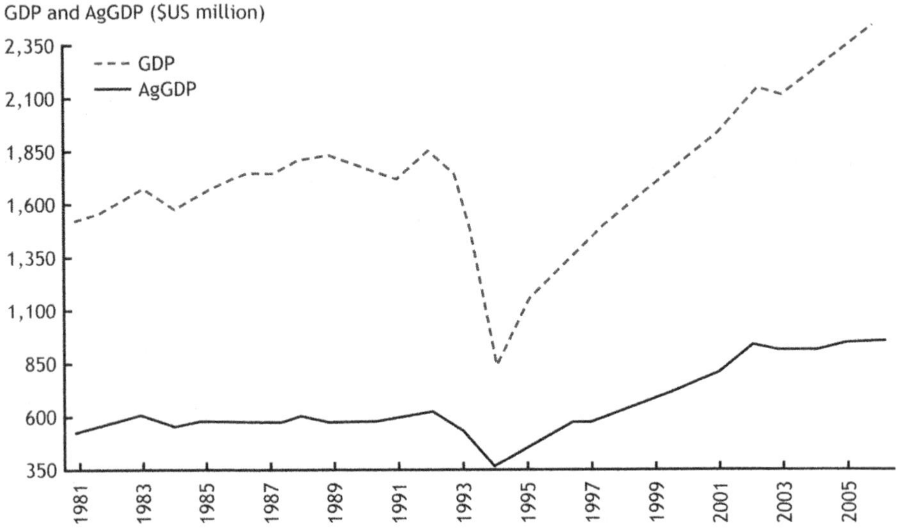

Source: World Bank (2008a).
Notes: Amounts are in constant 2000 U.S. dollars. AgGDP, agricultural gross domestic product; GDP, gross domestic product.

Figure A.2 Annual growth rate in GDP and AgGDP, 1981-2006

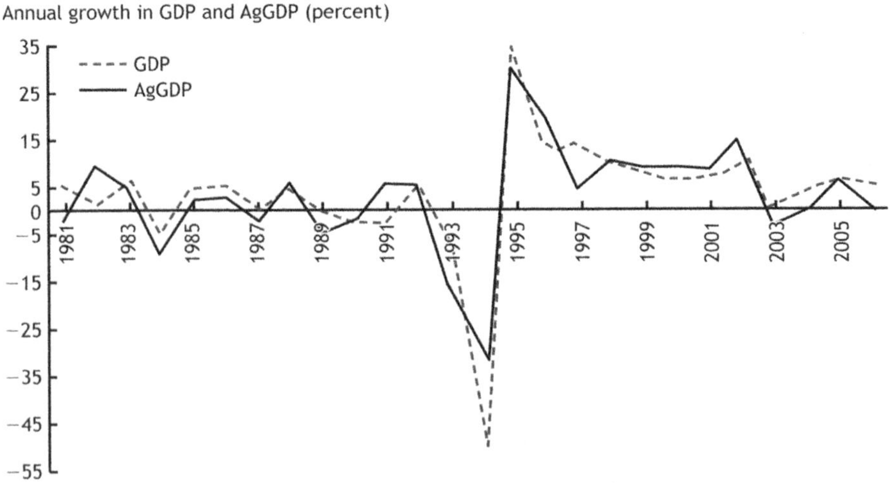

Source: World Bank (2008a).
Note: AgGDP, agricultural gross domestic product; GDP, gross domestic product.

Figure A.3 Selected crop production, first seasons, 1998-2008

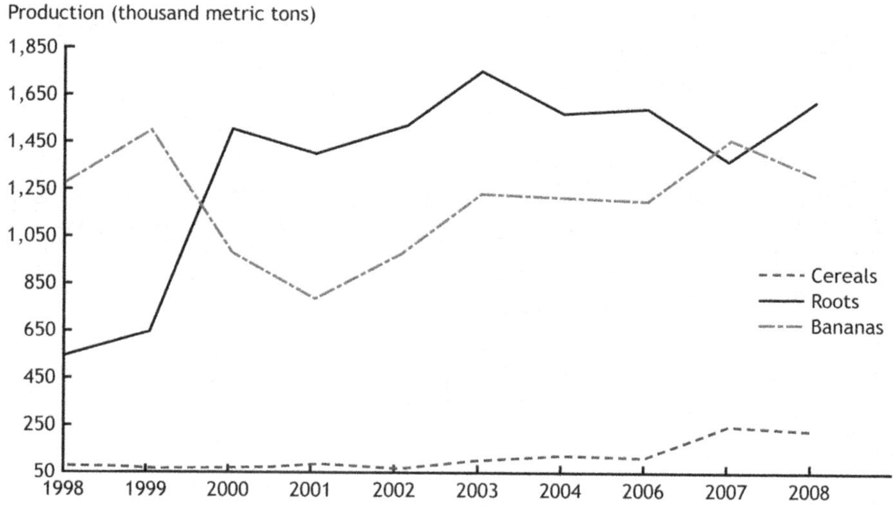

Source: Authors' calculations based on data from MINAGRI (2007).

Figure A.4 Selected crop production, second seasons, 1999-2007

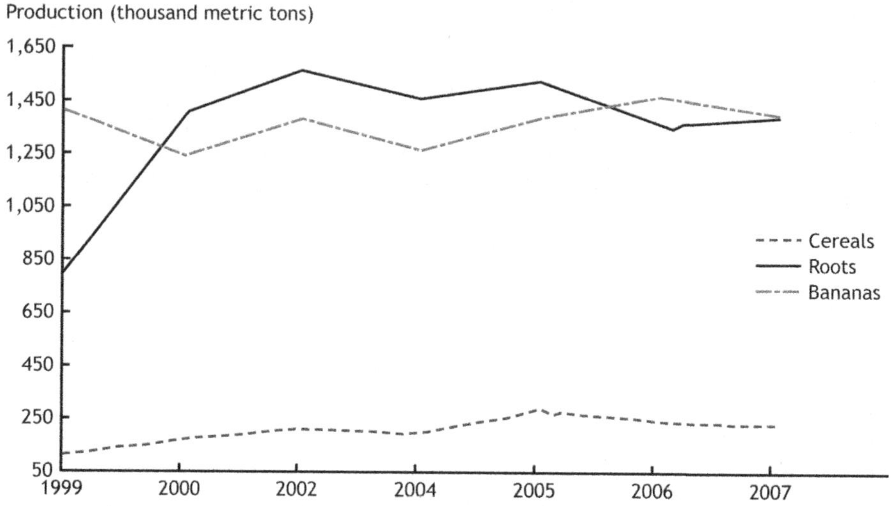

Source: Authors' calculations based on data from MINAGRI (2007).

Table A.1 Rwanda's economic structure: Agricultural and nonagricultural sectors

Sector/subsector	Share of GDP (%)	Share of employment (%)	Share of exports (%)	Export/ output ratio	Share of imports (%)	Import/ absorption ratio
Agriculture						
Grains						
Wheat	0.22	0.12	0.00	0.00	0.34	21.26
Maize	1.34	0.69	0.00	0.00	1.77	22.42
Paddy rice	0.79	0.39	0.00	0.00	2.79	36.38
Sorghum	1.77	0.95	0.00	0.00	0.00	0.00
Roots and tubers						
Irish potatoes	7.12	3.74	0.85	0.40	0.00	0.00
Sweetpotatoes	4.86	2.54	0.00	0.00	0.00	0.00
Cassava	3.52	1.88	0.00	0.00	0.00	0.00
Other roots	0.97	0.52	0.00	0.00	0.00	0.00
Other crops						
Pulses	4.77	2.48	0.00	0.00	0.00	0.00
Vegetables	0.99	0.64	0.18	0.62	0.00	0.00
Bananas	8.04	4.30	0.01	0.01	0.00	0.00
Fruits (excluding bananas)	1.09	0.71	0.05	0.15	0.00	0.00
Oilseeds	0.74	0.39	0.00	0.00	0.00	0.00
Export crops						
Coffee	0.95	0.71	0.00	0.00	0.00	0.00
Tea	0.53	0.33	0.00	0.00	0.00	0.00
Other export crops	0.41	0.19	0.41	4.13	0.02	1.10
Livestock						
Cattle	0.91	1.35	0.00	0.00	0.00	0.00
Sheep and goats	0.12	0.18	0.00	0.00	0.00	0.00
Swine	0.08	0.12	0.00	0.00	0.00	0.00
Poultry	0.10	0.14	0.00	0.00	0.00	0.00
Raw milk	0.53	0.79	0.00	0.00	0.00	0.00
Eggs	0.09	0.13	0.00	0.00	0.00	0.00
Other livestock	0.10	0.15	1.44	61.79	0.00	0.00
Forestry	2.65	3.63	0.00	0.00	0.00	0.00
Fishing	0.41	0.60	0.00	0.00	0.00	0.00
Nonagriculture						
Mining	0.78	0.85	23.74	100.00	0.00	0.00

Food processing						
Meat, fish, and dairy products	0.32	0.34	0.16	0.24	0.03	0.31
Processed cereals	0.36	0.44	0.03	0.08	4.36	43.20
Coffee processing	0.58	0.48	30.51	78.67	0.00	0.00
Tea processing	0.63	0.52	33.19	76.92	0.00	0.00
Other processing	0.08	0.09	0.00	0.00	9.24	86.58
Traditional beverages	1.14	1.37	0.00	0.00	0.00	0.00
Modern beverages	0.62	0.55	0.26	0.52	0.44	7.00
Tobacco	0.22	0.13	0.04	0.39	0.51	26.85
Other manufacturing						
Textile and clothing	0.31	0.44	0.20	0.79	7.86	71.46
Wood, paper, and printing	0.24	0.21	0.07	0.47	3.57	62.87
Chemicals	0.48	0.21	1.65	4.87	28.07	84.12
Nonmetallic minerals	0.64	0.30	0.23	0.69	2.61	35.10
Other manufacturing	0.81	0.54	6.99	20.47	38.40	88.03
Electricity, gas, and water	0.64	0.53	0.00	0.00	0.00	0.00
Construction	6.60	8.50	0.00	0.00	0.00	0.00
Services						
Wholesale and retail	9.24	11.13	0.00	0.00	0.00	0.00
Hotels and restaurants	0.82	0.93	0.00	0.00	0.00	0.00
Transport	3.53	4.80	0.00	0.00	0.00	0.00
Communications	2.25	2.08	0.00	0.00	0.00	0.00
Finance and insurance	5.07	6.94	0.00	0.00	0.00	0.00
Real estate	6.11	2.11	0.00	0.00	0.00	0.00
Business services	1.79	2.22	0.00	0.00	0.00	0.00
Repair	0.66	0.59	0.00	0.00	0.00	0.00
Public administration	6.37	14.37	0.00	0.00	0.00	0.00
Education	4.26	7.47	0.00	0.00	0.00	0.00
Health	1.38	2.21	0.00	0.00	0.00	0.00
Other personal services	0.98	1.95	0.00	0.00	0.00	0.00
Total for agriculture	43.10	27.70	2.94	0.27	4.91	2.00
Total for nonagriculture	56.90	72.30	97.06	4.25	95.09	21.51
Industry	14.45	15.50	0.00	0.00	0.00	0.00
Manufacturing	6.43	5.62	0.00	0.00	0.00	0.00
Other industry	8.02	9.88	0.00	0.00	0.00	0.00
Services	42.45	56.80	0.00	0.00	0.00	0.00
Total	100.00	100.00	100.00	2.97	100.00	15.75

Source: Arnault Emini (2007).
Note: GDP, gross domestic product.

Table A.2 Per capita total expenditure and average budget share by major commodity, 1999-2001

Subsector	Rural quintile					Rural average	Urban quintile					Urban average	National average
	Lowest	Second	Third	Fourth	Highest		Lowest	Second	Third	Fourth	Highest		
Average budget share (%)													
Maize	5.2	4.5	4.6	4.1	2.5	3.8	1.3	1.2	1.5	1.1	0.6	0.6	2.9
Rice	0.7	0.7	1.1	1.4	2.4	1.5	1.4	4.8	5.8	4.6	4.0	4.1	2.2
Sorghum	3.5	3.0	3.1	3.1	2.2	2.8	2.2	2.4	1.5	1.7	0.7	0.8	2.3
Wheat	0.3	0.4	0.3	0.4	0.8	0.5	0.3	0.5	0.7	1.1	2.0	1.9	0.9
Cassava	8.3	6.9	6.2	5.3	3.4	5.3	6.6	4.8	6.3	5.0	1.7	2.0	4.4
Irish potatoes	7.6	7.9	7.6	6.3	4.8	6.4	12.6	9.5	9.5	7.1	3.1	3.5	5.6
Sweetpotatoes	16.8	19.0	17.5	14.1	8.5	13.8	7.4	5.3	4.6	3.0	0.7	1.0	10.2
Other roots	2.3	2.8	2.9	3.4	2.0	2.7	0.5	0.3	0.5	0.4	0.1	0.2	2.0
Pulses	14.8	14.3	12.5	11.4	7.3	10.9	11.4	8.5	8.6	5.4	2.2	2.6	8.6
Bananas	2.1	3.5	4.3	4.8	5.9	4.7	0.5	3.0	2.3	2.8	2.3	2.3	4.0
Vegetables	12.8	10.3	8.5	6.6	5.5	7.6	5.7	10.1	6.5	7.0	5.5	5.6	7.0
Fruits (excluding bananas)	3.6	2.6	2.7	2.1	2.1	2.4	2.5	1.8	1.4	2.4	1.3	1.3	2.1
Oilseeds	2.4	2.7	3.0	3.4	3.5	3.2	3.9	4.1	4.2	4.5	3.8	3.9	3.4
Other crops	0.4	0.6	0.9	1.3	2.3	1.4	3.2	5.1	4.6	5.0	3.6	3.7	2.0
Livestock	1.7	1.9	2.5	4.0	6.4	4.0	2.2	5.4	4.5	7.3	9.4	9.1	5.5
Processed food	3.3	4.0	5.0	5.6	8.0	5.9	3.7	3.1	3.2	3.6	4.5	4.4	5.5
Nonfood	14.2	14.9	17.3	22.7	32.6	23.2	34.6	30.3	34.1	38.0	54.5	52.8	31.5
Total expenditure per capita (RWF)	15,428	27,104	38,438	54,633	102,759	46,016	14,802	29,324	40,806	58,727	204,151	165,276	57,824

Source: Authors' calculations using data from MINECOFIN (2003).
Note: RWF, Rwandan franc.

Table A.3 Per capita total expenditure and average budget share by major commodity, 2005-06

Subsector	Rural quintile					Rural average	Urban quintile					Urban average	National average
	Lowest	Second	Third	Fourth	Highest		Lowest	Second	Third	Fourth	Highest		
Average budget share (%)													
Maize	3.8	3.4	4.1	3.2	3.4	3.5	3.9	4.7	4.1	3.4	1.0	1.4	2.8
Rice	0.6	1.1	1.3	1.5	2.1	1.6	1.0	1.6	1.7	2.8	3.2	3.0	2.1
Sorghum	1.7	2.4	2.5	2.1	1.9	2.1	1.2	2.2	2.0	1.7	0.5	0.7	1.6
Wheat	0.1	0.3	0.1	0.2	0.1	0.1	0.0	0.2	0.0	0.0	0.1	0.1	0.1
Cassava	6.6	7.3	5.9	4.9	2.9	4.8	8.1	9.5	8.4	6.3	1.7	2.5	4.1
Irish potatoes	6.2	7.2	7.0	5.8	4.4	5.7	6.4	5.1	6.1	5.9	2.9	3.4	4.9
Sweetpotatoes	17.3	12.6	10.2	8.0	6.3	9.2	10.1	7.0	6.0	4.7	0.9	1.7	6.7
Other roots	0.8	0.7	0.8	0.9	0.3	0.6	0.1	0.4	0.3	0.3	0.1	0.1	0.5
Pulses	18.5	16.4	14.7	12.0	9.6	12.8	14.4	14.3	12.7	8.2	2.5	3.8	9.7
Bananas	6.2	5.9	6.5	8.2	6.8	6.9	3.2	3.8	4.8	4.8	2.2	2.5	5.5
Vegetables	4.5	4.0	3.9	3.8	3.2	3.7	5.5	5.3	4.9	4.3	3.0	3.3	3.6
Fruits (excluding bananas)	0.9	0.9	1.0	0.9	0.8	0.9	1.0	1.0	1.1	1.0	0.8	0.9	0.9
Oilseeds	2.4	2.6	2.9	3.1	3.2	3.0	3.7	3.5	3.4	3.5	2.5	2.6	2.9
Other crops	0.3	0.6	0.7	1.1	1.5	1.1	0.5	1.3	2.1	2.4	2.4	2.3	1.5
Livestock	1.3	2.3	3.0	3.7	5.0	3.7	2.4	2.8	4.0	4.8	7.3	6.8	4.7
Processed food	4.8	6.0	7.2	8.2	9.5	7.9	5.1	5.3	6.8	7.7	7.7	7.6	7.8
Nonfood	24.1	26.3	28.1	32.3	38.8	32.4	33.3	32.0	32.0	38.0	61.3	57.2	40.7
Total expenditure per capita (RWF)	30,724	51,390	71,727	104,948	230,150	90,039	28,713	50,860	68,603	106,318	361,480	229,310	112,942

Source: Authors' calculations based on data from MINECOFIN (2007a).
Note: RWF, Rwandan franc.

Table A.4 Marginal budget share by major commodity, 1999-2001

Subsector	Rural quintile					Rural average	Urban quintile					Urban average	National average
	Lowest	Second	Third	Fourth	Highest		Lowest	Second	Third	Fourth	Highest		
Maize	2.3	1.7	1.4	1.0	0.5	1.4	2.5	1.7	1.3	1.0	-0.1	0.2	1.3
Rice	8.0	6.8	6.1	5.4	4.3	6.2	8.3	6.7	6.0	5.3	3.1	3.8	5.9
Sorghum	2.3	1.8	1.5	1.2	0.7	1.5	2.4	1.7	1.4	1.1	0.2	0.5	1.4
Wheat	2.4	2.3	2.3	2.2	2.1	2.3	2.5	23.	2.2	2.2	2.0	2.0	2.2
Cassava	3.5	2.8	2.4	1.9	1.2	2.4	3.7	2.7	2.3	1.9	0.5	1.0	2.3
Irish potatoes	6.4	5.2	4.4	3.8	2.6	4.5	6.7	5.0	4.4	3.6	1.5	2.1	4.3
Sweetpotatoes	3.7	2.5	1.8	1.2	0.1	1.9	4.0	2.4	1.8	1.1	-1.0	-0.3	1.7
Other roots	1.4	1.0	0.8	0.6	0.2	0.8	1.5	1.0	0.8	—	—	—	—
Pulses	5.4	4.1	3.4	2.7	1.5	3.4	5.7	4.0	3.3	2.5	0.3	1.0	3.2
Bananas	6.4	5.2	4.6	4.0	3.0	4.7	6.6	5.1	4.5	3.9	1.9	2.5	4.5
Vegetables	6.6	5.8	5.4	5.0	4.3	5.5	6.8	5.8	5.4	4.9	3.6	4.0	5.3
Fruits (excluding bananas)	1.6	1.4	1.4	1.3	1.2	1.4	1.6	1.4	1.3	1.3	1.0	1.1	1.3
Oilseeds	7.1	6.1	5.5	4.9	4.0	5.5	7.3	6.0	5.4	4.8	3.0	3.6	5.3
Other crops	5.8	5.1	4.7	4.2	3.5	4.7	6.0	5.0	4.6	4.2	2.9	3.3	4.6
Livestock	13.8	12.7	12.1	11.5	10.5	12.1	14.0	12.6	12.0	11.4	9.5	10.1	11.9
Processed food	8.1	7.2	6.7	6.2	5.3	6.7	8.3	7.1	6.6	6.1	4.5	5.0	6.5
Nonfood	15.1	28.3	35.7	43.0	55.1	35.0	12.1	29.4	36.7	44.3	67.1	60.1	37.5

Source: Authors' calculations based on data from MINECOFIN (2003).

Table A.5 Marginal budget share by major commodity, 2005-06 (percent)

	Rural quintile						Urban quintile						
Subsector	Lowest	Second	Third	Fourth	Highest	Rural average	Lowest	Second	Third	Fourth	Highest	Urban average	National average
Maize	3.1	2.6	2.3	2.0	1.4	2.3	3.1	2.6	2.3	2.0	1.1	1.7	2.2
Rice	5.7	4.9	4.5	4.0	3.2	4.5	5.7	5.0	4.6	4.0	2.7	3.6	4.4
Sorghum	1.5	1.2	1.0	0.9	0.6	1.1	1.5	1.2	1.1	0.9	0.4	0.7	1.0
Wheat	0.1	0.1	0.1	0.1	0.1	0.1	0.1	0.1	0.1	0.1	0.1	0.1	0.1
Cassava	2.1	1.7	1.5	1.3	0.8	1.5	2.1	1.7	1.5	1.3	0.6	1.0	1.4
Irish potatoes	4.7	4.0	3.6	3.2	2.4	3.6	4.7	4.0	3.7	3.2	2.0	2.8	3.5
Sweetpotatoes	2.8	2.4	2.2	2.0	1.5	2.2	2.8	2.4	2.2	1.9	1.3	1.7	2.1
Other roots	0.3	0.2	0.2	0.1	0.1	0.2	0.3	0.2	0.2	0.1	0.1	0.1	0.2
Pulses	5.7	4.8	4.3	3.7	2.7	4.3	5.7	4.8	4.4	3.7	2.2	3.2	4.2
Bananas	6.7	5.6	5.0	4.3	3.0	5.0	6.7	5.6	5.0	4.2	2.4	3.6	4.8
Vegetables	5.2	4.6	4.2	3.8	3.1	4.2	5.2	4.6	4.2	3.8	2.7	3.4	4.1
Fruits (excluding bananas)	1.4	1.2	1.1	1.0	0.9	1.2	1.4	1.2	1.2	1.0	0.8	1.0	1.1
Oilseeds	4.4	3.8	3.5	3.1	2.3	3.5	4.5	3.8	3.5	3.0	2.0	2.7	3.4
Other crops	4.1	3.6	3.2	2.9	2.2	3.3	4.1	3.6	3.3	2.9	1.9	2.5	3.1
Livestock	13.2	11.8	10.9	10.0	8.4	11.0	13.2	11.8	11.0	10.0	7.6	9.2	10.7
Processed food	12.5	11.3	10.6	9.8	8.5	10.7	12.5	11.3	10.7	9.8	7.8	9.1	10.4
Nonfood	26.6	36.1	41.7	47.9	58.8	41.3	26.4	35.9	41.1	48.1	64.4	53.5	43.3

Source: Authors' calculations based on data from MINECOFIN (2007a).

Table A.6 Annual growth rates assumed for sector total factor productivity in model simulations, by scenario (percent)

Scenario	Baseline run	S1	S2	S3	S4	S20	S21
Maize	2.3	2.7	–	–	2.7	2.7	2.7
Wheat	3.0	–	2.0	–	2.3	2.3	2.3
Paddy rice	1.0	–	–	2.3	2.0	2.0	2.0
Sorghum	2.5	–	–	–	0.8	0.8	0.8
Scenario	**Baseline run**	**S5**	**S6**	**S7**	**S8**	**S20**	**S21**
Irish potatoes	1.8	2.5	–	–	2.5	2.5	2.5
Sweetpotatoes	1.3	–	2.6	–	2.6	2.6	2.6
Cassava	1.8	–	–	2.7	2.7	2.7	2.7
Other roots	1.4	–	–	–	8.6	8.6	8.6
Scenario	**Baseline run**	**S9**	**S10**	**S11**	**–**	**S20**	**S21**
Pulses	1.4	1.2	–	–	–	1.2	1.2
Bananas	2.5	–	1.4	–	–	1.4	1.4
Oilseeds	2.1	–	–	2.1	–	2.1	2.1
Scenario	**Baseline run**	**S12**	**S13**	**S14**	**S15**	**S20**	**S21**
Coffee	3.3	4.3	–	–	4.3	–	4.3
Tea	3.3	–	3.8	–	3.8	–	3.8
Vegetables	2.0	–	–	0.8	0.8	–	0.8
Fruits (excluding bananas)	2.5	–	–	0.5	0.5	–	0.5
Other export crops	2.2	–	–	3.2	3.2	–	3.2
Scenario	**Baseline run**	**S16**	**S17**	**S18**	**S19**	**S20**	**S21**
Poultry	3.0	5.0	–	–	5.0	5.0	5.0
Eggs	3.3	6.7	–	–	6.7	6.7	6.7
Cattle	3.0	–	2.6	–	2.6	2.6	2.6
Sheep and goats	3.0	–	2.9	–	2.9	2.9	2.9
Swine	3.0	–	2.6	–	2.6	2.6	2.6
Raw milk	3.3	–	7.0	–	7.0	7.0	7.0
Other livestock	3.3	–	6.4	–	6.4	6.4	6.4
Fishing	1.8	–	–	3.7	3.7	–	3.7
Forestry	1.8	–	–	–	–	–	5.3

Source: Authors.

Notes: –, not applicable; S*nn*, Scenario *nn*.

Table A.7 Annual growth rates assumed for sector area in simulations (percent)

		Additional annual growth rate					
	Baseline run	**Scenario 3**	**Scenario 4**	**Scenario 12**	**Scenario 13**	**Scenario 14**	**Scenario 15**
Rice	3.0	5.8	5.8	–	–	–	–
Export crops							
Coffee	2.0	–	–	2.0	–	–	2.0
Tea	2.0	–	–	–	2.0	–	2.0
Other export crops	1.2	–	–	–	–	2.4	2.4

Source: Authors.
Note: –, not applicable.

APPENDIX B

Mathematical Presentation of the DCGE Model of Rwanda

Table B.1 Computable general equilibrium model: Sets and parameters

Symbol	Explanation
Sets	
$a \in A$	Activities
$a \in ALEO(\subset A)$	Activities with a Leontief function at the top of the technology next
$c \in C$	Commodities
$CD(\subset C)$	Subset of commodities with domestic sales of domestic output
$CDN(\subset C)$	Subset of commodities not in *CD*
$CE(\subset C)$	Subset of exported commodities
$CEN(\subset C)$	Subset of commodities not in *CE*
$CM(\subset C)$	Subset of aggregate imported commodities
$CMN(\subset C)$	Subset of commodities not in *CM*
$CT(\subset C)$	Subset of transaction service commodities
$CX(\subset C)$	Subset of commodities with domestic production
$f \in F$	Factors
gov	Government
$h \in H(\subset INSDNG)$	Households
$i \in INS$	Institutions (domestic and rest of world)
$INSD(\subset INS)$	Subset of domestic institutions
$INSDNG(\subset INSD)$	Subset of domestic nongovernmental institutions
$r \in R$	Subnational regions (provinces)
$z \in Z$	Districts or zones within a region
Parameters	
$cwts_c$	Weight of commodity *c* in the consumer price index
$dwts_c$	Weight of commodity *c* in the producer price index

Table B.1 Continued

Symbol	Explanation
ica_{ca}	Quantity of commodity c as intermediate input per unit of activity a
$icd_{cc'}$	Quantity of commodity c as trade input per unit of c' produced and sold domestically
$ice_{cc'}$	Quantity of commodity c as trade input per exported unit of c'
$icm_{cc'}$	Quantity of commodity c as trade input per imported unit of commodity c'
$inta_a$	Quantity of aggregate intermediate input per unit of activity a
iva_a	Quantity of aggregate intermediate input per unit of value-added activity a
$\overline{mps}_i$	Base savings rate for domestic institution i
mps_i	Flexible savings rate for domestic institution i
$qdst_c$	Quantity of stock change of commodity c
$\overline{qg}_c$	Base-year quantity of government demand of commodity c
$\overline{qinv}_c$	Baseline-year quantity of private investment demand for commodity c
$shif_{if}$	Share for domestic institution i in income of factor f
$shii_{ii'}$	Share of net income of institution i' to institution i ($i' \in$ INSDNG′; $i \in$ INSDNG)
ta_a	Tax rate for activity a
$\overline{tins}_i$	Exogenous direct tax rate for domestic institution i
$tins01_i$	0-1 parameter equal to 1 for institution i with potentially flexed direct tax rates
tm_{cr}	Import tariff rate for commodity c in region r
tq_c	Rate of sales tax for commodity c
$trnsfr_{if}$	Transfer from factor f to institution i

Table B.2 Computable general equilibrium model: Elasticities, coefficients, and exogenous variables

Symbol	Explanation
Elasticities and shift coefficients	
α_a^a	Efficiency parameter in the constant elasticity-of-substitution activity function for activity a
α_a^{va}	Efficiency parameter in the constant elasticity-of-substitution value-added function for activity a

(*continued*)

Table B.2 Continued

Symbol	Explanation
α_{fa}^{vaf}	Efficiency parameter of factor f in value-added function for activity a
α_{c}^{ac}	Shift parameter for the domestic commodity aggregation function for commodity c
α_{c}^{q}	Armington function shift parameter
α_{c}^{t}	Shift parameter in the constant elasticity-of-transformation function for commodity c
β^{a}	Sector capital mobility factor
β_{ch}^{m}	Marginal share of consumption spending on marketed commodity c for household h
δ_{a}^{a}	Share parameter in the constant elasticity-of-transformation activity function for activity a
δ_{ac}^{ac}	Share parameter in the domestic commodity aggregation function for activity a and commodity c
δ_{cr}^{q}	Share parameter in the Armington function for commodity c in region r
δ_{cr}^{t}	Share parameter in the constant elasticity-of-transformation function for commodity c in region r
δ_{fa}^{va}	Share parameter in the constant elasticity-of-substitution value-added function for factor f in activity a
υ_{f}	Capital depreciation rate for factor f
γ_{ch}^{m}	Subsistence consumption of marketed commodity c for household h
θ_{ac}	Yield of output for commodity c per unit of activity a
ρ_{a}^{a}	Exponent in the constant elasticity-of-substitution production function for activity a
ρ_{a}^{va}	Exponent in the constant elasticity-of-substitution value-added function for activity a
ρ_{c}^{ac}	Exponent in the domestic commodity aggregation function for commodity c
ρ_{c}^{q}	Exponent in the Armington function for commodity c
ρ_{c}^{t}	Exponent in the constant elasticity-of-transformation function for commodity c
η_{fat}^{a}	Sector share of new capital for factor f and activity a in time period t
Exogenous Variables	
$\overline{CPI}$	Consumer price index
$\overline{DTINS}$	Change in domestic institution tax share (= 0 for baseline)
$\overline{FSAV}$	Foreign savings (foreign currency unit)
$\overline{GADJ}$	Government consumption adjustment factor
$\overline{IADJ}$	Investment adjustment factor
$\overline{MPSADJ}$	Scaling factor for savings rate (= 0 for baseline)

Table B.2 Continued

Symbol	Explanation
pwe_{cr}	Export price for commodity *c* in region *r* (foreign currency)
pwm_{cr}	Import price for commodity *c* in region *r* (foreign currency)
QFS_{ft}	Quantity supplied of factor *f* in time period *t*
$\overline{TINSADJ}$	Direct tax scaling factor (= 0 for baseline)
$\overline{WFDIST}_{fa}$	Wage distortion factor for factor *f* in activity *a*

Table B.3 Computable general equilibrium model: Endogenous variables

Symbol	Explanation
AWF^{a}_{ft}	Average capital rental rate for factor *f* in time period *t*
$DMPS$	Change in domestic institution savings rates
DPI	Producer price index for domestically marketed output
EG	Government expenditures
EH_h	Consumption spending for household *h*
EXR	Exchange rate (local currency unit per foreign currency unit)
$GSAV$	Government savings
MPS_i	Marginal propensity to save for domestic nongovernmental institution *i*
PA_a	Price for activity *a* (unit gross revenue)
PDD_c	Demand price for commodity *c* produced and sold domestically
PDS_c	Supply price for commodity *c* produced and sold domestically
PE_{cr}	Export price for commodity *c* in region *r* (domestic currency)
$PINTA_a$	Aggregate intermediate input price for activity *a*
PK_{ft}	Unit price of capital for factor *f* in time period *t*
PM_{cr}	Import price for commodity *c* in region *r* (domestic currency)
PT_{ct}	Composite price of commodity *c* in time period *t*
PVA_a	Value-added price of activity *a* (factor income per unit of activity)
PX_c	Aggregate producer price of commodity *c*
$PXAC_{ac}$	Producer price of commodity *c* for activity *a*

(continued)

Table B.3 Continued

Symbol	Explanation
QA_a	Quantity (level) of activity a
QD_c	Quantity of commodity c (from domestic output) sold domestically
QE_{cr}	Quantity of exports for commodity c in region r
QF_{fat}	Quantity of demand for factor f in activity a in time period t
QT_c	Government consumption demand for commodity c
QH_{ch}	Quantity of commodity c consumed by household h
QHA_{ach}	Quantity of commodity c consumed by household h for activity a
$QINT_{ca}$	Quantity of commodity c used as intermediate input to activity a
$QINTA_a$	Quantity of aggregate intermediate input for activity a
$QINV_{ct}$	Quantity of investment demand for commodity c in time period t
QM_{cr}	Quantity of imports of commodity c in region r
QQ_c	Quantity of commodity c supplied to domestic market (composite supply)
QT_c	Quantity of commodity c demanded as trade input
QVA_a	Quantity of (aggregate) value added for activity a
QX_c	Aggregated quantity of domestic output of commodity c
$QXAC_{ac}$	Quantity of output of commodity c from activity a
RWF_f	Real average price for factor f
$TABS$	Total nominal absorption
$TINS_i$	Direct tax rate for institution i ($i \in$ INSDNG)
$TRII_{ii'}$	Transfers from institution i' to institution i (i, $i' \in$ INSDNG)
WF_{ft}	Average price of factor f in time period t
$WFDIST_{fat}$	Wage distortion factor for factor f in activity a in time period t
YF_{if}	Income for institution i from factor f
YG	Government revenue
YI_i	Income of domestic nongovernmental institution i
YIF_{if}	Income of domestic institution i from factor f
ΔK^a_{fat}	Quantity of new capital for factor f and activity a for time period t

Note: INSDNG, domestic nongovernmental institution (see Table B.1).

Table B4 Computable general equilibrium model: Equations

Production and price

$$QINT_{ca} = ica_{ca} \cdot QINTA_{a} \quad (1)$$

$$PINTA_{a} = \sum_{c \in C} PQ_{c} \cdot ica_{ca} \quad (2)$$

$$QVA_{a} = \alpha_{a}^{va} \cdot \left(\sum_{f \in F} \delta_{fa}^{va} \cdot (\alpha_{fa}^{vaf} \cdot QF_{fa})^{-\rho_{a}^{va}} \right)^{-\frac{1}{\rho_{a}^{va}}} \quad (3)$$

$$W_{f} \cdot \overline{WFDIST}_{fa} = PVA_{a} \cdot QVA_{a} \cdot \left(\sum_{f \in F'} \delta_{fa}^{va} \cdot (\alpha_{fa}^{vaf} \cdot QF_{fa})^{-\rho_{a}^{va}} \right)^{-1} \cdot \delta_{fa}^{va} \cdot (\alpha_{fa}^{vaf} \cdot QF_{fa})^{-\rho_{a}^{va}-1} \quad (4)$$

$$QVA_{a} = iva_{a} \cdot QA_{a} \quad (5)$$

$$QINTA_{a} = inta_{a} \cdot QA_{a} \quad (6)$$

$$PA_{a} \cdot (1-ta_{a}) \cdot QA_{a} = PVA_{a} \cdot QVA_{a} + PINTA_{a} \cdot QINTA_{a} \quad (7)$$

$$QXAC_{ac} = \theta_{ac} \cdot QA_{a} \quad (8)$$

$$PA_{a} = \sum_{c \in C} PXAC_{ac} \cdot \theta_{ac} \quad (9)$$

$$QX_{c} = \alpha_{c}^{ac} \cdot \left(\sum_{a \in A} \delta_{ac}^{ac} \cdot QXAC_{ac}^{-\rho_{c}^{ac}} \right)^{-\frac{1}{\rho_{c}^{ac}-1}} \quad (10)$$

$$PXAC_{ac} = PX_{c} \cdot QX_{c} \left(\sum_{a \in A'} \delta_{ac}^{ac} \cdot QXAC_{ac}^{-\rho_{c}^{ac}} \right)^{-1} \cdot \delta_{ac}^{ac} \cdot QXAC_{ac}^{-\rho_{c}^{ac}-1} \quad (11)$$

$$PE_{cr} = pwe_{cr} \cdot EXR - \sum_{c' \in CT} PQ_{c} \cdot ice_{c'c} \quad (12)$$

$$QX_{c} = \alpha_{c}^{t} \cdot \left(\sum_{r \in R} \delta_{cr}^{t} \cdot QE_{cr}^{\rho_{c}^{t}} + (1 - \sum_{r \in R} \delta_{cr}^{t}) \cdot QD_{c}^{\rho_{c}^{t}} \right)^{\frac{1}{\rho_{c}^{t}}} \quad (13)$$

(continued)

$$\frac{QE_{cr}}{QD_c} = \left(\frac{PE_{cr}}{PDS_c} \cdot \frac{1 - \sum_{r \in R} \delta^t_{cr}}{\delta^t_{cr}} \right)^{\frac{1}{\rho^t_c - 1}} \tag{14}$$

$$QX_c = QD_c + \sum_{r \in R} QE_{cr} \tag{15}$$

$$PX_c \cdot QX_c = PDS_c \cdot QD_c + \sum_{r \in R} PE_{cr} \cdot QE_{cr} \tag{16}$$

$$PDD_c = PDS_c + \sum_{c' \in CT} PQ_{c'} \cdot icd_{c'c} \tag{17}$$

$$PM_{cr} = pwm_{cr} \cdot (1 + tm_{cr}) \cdot EXR + \sum_{c' \in CT} PQ_{c'} \cdot icm_{c'c} \tag{18}$$

$$QQ_c = \alpha^q_c \cdot \left(\sum_{r \in R} \delta^q_{cr} \cdot QM_{cr}^{-\rho^q_c} + (1 - \sum_{r \in R} \delta^q_{cr}) \cdot QD_c^{-\rho^q_c} \right)^{-\frac{1}{\rho^q_c}} \tag{19}$$

$$\frac{QM_{cr}}{QD_c} = \left(\frac{PDD_c}{PM_c} \cdot \frac{\delta^q_c}{1 - \sum_{r \in R} \delta^q_{cr}} \right)^{\frac{1}{1 + \rho^q_c}} \tag{20}$$

$$QQ_c = QD_c + \sum_{r \in R} QM_{cr} \tag{21}$$

$$PQ_c \cdot (1 - tq_c) \cdot QQ_c = PDD_c \cdot QD_c + \sum_{r \in R} PM_{cr} \cdot QM_{cr} \tag{22}$$

$$QT_c = \sum_{c' \in C'} (icm_{cc'} \cdot QM_{c'} + ice_{cc'} \cdot QE_{c'} + icd_{cc'} \cdot QD_{c'}) \tag{23}$$

$$\overline{CPI} = \sum_{c \in C} PQ_c \cdot cwts_c \tag{24}$$

$$DPI = \sum_{c \in C} PDS_c \cdot dwts_c \tag{25}$$

Institutional incomes and domestic demand

$$YF_f = \sum_{a \in A} WF_f \cdot \overline{WFDIST}_{fa} \cdot QF_{fa} \tag{26}$$

$$YIF_{if} = shif_{if} \cdot [YF_f - trnsfr_{row\,f} \cdot EXR] \tag{27}$$

$$YI_i = \sum_{f \in F} YIF_{if} + \sum_{i' \in INSDNG'} TRII_{ii'} + trnsfr_{i\,gov} \cdot \overline{CPI} + trnsfr_{i\,row} \cdot EXR \tag{28}$$

$$TRII_{ii'} = shii_{ii'} \cdot (1 - MPS_{i'}) \cdot (1 - \overline{tins}_{i'}) \cdot YI_{i'} \tag{29}$$

$$EH_h = \left(1 - \sum_{i' \in INSDNG} shii_{ih}\right) \cdot (1 - MPS_h) \cdot (1 - \overline{tins}_h) \cdot YI_h \tag{30}$$

$$PQ_c = QH_{ch} = PQ_c \cdot \gamma^m_{ch} + \beta^m_{ch} \cdot \left(EH_h - \sum_{c' \in C} PQ_{c'} \cdot \gamma^m_{c'h}\right) \tag{31}$$

$$QNIV_c = \overline{IADJ} \cdot \overline{qniv}_c \tag{32}$$

$$QG_c = \overline{GADJ} \cdot \overline{qg}_c \tag{33}$$

$$EG = \sum_{c \in C} PQ_c \cdot GQ_c + \sum_{i \in INSDNG} trnsfr_{i\,gov} \cdot \overline{CPI} \tag{34}$$

System constraints and macroeconomic closures

$$YG = \sum_{i \in INSDNG} \overline{tins}_i \cdot YI_i + \sum_{c \in CMN} tm_c \cdot pwm_c \cdot QM_c \cdot EXR + \sum_{c \in C} tq_c \cdot PQ_c \cdot QQ_c + \sum_{f \in F} YF_{gov\,f} + trnsfr_{gov\,row} \cdot EXR \tag{35}$$

$$QQ_c = \sum_{a \in A} QINT_{ca} + \sum_{h \in H} QH_{ch} + QC_c + QINV_c + qdst_c + QT_c \tag{36}$$

$$\sum_{a \in A} QF_{fa} = \overline{QFS}_f \tag{37}$$

(continued)

$$YG = EG + GSAV \tag{38}$$

$$\sum_{rc \in CMN} pwm_{cr} \cdot QM_{cr} + \sum_{f \in F} trnsfr_{row\,f} = \sum_{rc \in CEN} pwe_{cr} \cdot QE_{cr} + \sum_{i \in INSD} trnsfr_{row\,i} + \overline{FSAV} \tag{39}$$

$$\sum_{i \in INSDNG} MPS_i \cdot (1 - \overline{tins_i}) \cdot YI + GSAV + EXR\,\overline{FSAV} = \sum_{c \in C} PQ_c\, QINV_c + \sum_{c \in C} PQ_c\, qdst_c \tag{40}$$

$$MPS_i = \overline{mps_i} \cdot (1 + \overline{MPSADJ}) \tag{41}$$

Capital accumulation and allocation

$$AFW^a_{ft} = \sum_{a \in A} \left[\left(\frac{QF_{fat}}{\sum_{a'} QF_{fa't}} \right) \cdot WF_{ft} \cdot WFDIST_{fat} \right] \tag{42}$$

$$\eta^a_{fat} = \left(\frac{QF_{fat}}{\sum_{a \in A} QF_{fa't}} \right) \cdot \left(\beta^a \cdot \left(\frac{WF_{ft} \cdot WFDIST_{fat}}{AWF^a_{ft}} - 1 \right) + 1 \right) \tag{43}$$

$$\Delta K^a_{fat} = \eta^a_{fat} \cdot \left(\frac{\sum_{a \in A} PQ_{ct} \cdot QINV_{ct}}{PK_{ft}} \right) \tag{44}$$

$$PK_{ft} = \sum_{c \in C} PQ_{ct} \cdot \frac{QINV_{ct}}{\sum_{c' \in C'} QNIV_{c't}} \tag{45}$$

$$QF_{fa(t+1)} = QF_{fat} \cdot \left(1 + \frac{\Delta K^a_{fat}}{QF_{fat}} - \upsilon_f \right) \tag{46}$$

$$QFS_{f(t+1)} = QFS_{ft} \cdot \left(1 + \frac{\sum_{a \in A} \Delta K^a_{fat}}{QFS_{ft}} - \upsilon_f \right) \tag{47}$$

Note: INSDNG, domestic nongovernmental institutions; *row*, rest of world.

Sensitivity Test of the DCGE Model

Table C.1 Sensitivity test: Price comparison

	Substitution elasticity 25% higher than in report			Substitution elasticity 25% lower than in report	
Subsector	Price, 2015	Difference from model result (%)	Model results in report	Price, 2015	Difference from model result (%)
Maize	1.04	-1.27	1.05	1.07	1.91
Rice	0.77	2.27	0.75	0.73	-3.24
Irish potatoes	1.32	-0.22	1.32	1.33	0.26
Vegetables	0.82	-0.22	0.82	0.82	0.27
Beef	0.75	-0.25	0.75	0.75	0.30
Poultry	0.58	-0.22	0.58	0.58	0.27

Source: Numbers are model simulation results derived from authors' calculations using 2006 data from MINAGRI (2007).

Notes: Substitution elasticities for imports are those used in the Armington function; substitution elasticities for exports are those used in constant elasticity of transformation (CET) functions. Prices are normalized to 1 in the first year.

Table C.2 Substitution elasticities used in the model

Subsector	Elasticity for import substitution	Elasticity for export substitution
Grains		
Wheat	6.5	6.5
Maize	8.0	8.0
Paddy rice	6.5	6.5
Sorghum	6.0	6.0
Roots and tubers		
Irish potatoes	6.0	6.0
Sweetpotatoes	6.0	6.0
Cassava	6.0	6.0
Other roots	6.0	6.0

(continued)

Table C.2 Continued

Subsector	Elasticity for import substitution	Elasticity for export substitution
Other crops		
Pulses	6.0	6.0
Vegetables	6.0	8.0
Bananas	6.0	6.0
Fruits (excluding bananas)	6.0	6.0
Oilseeds	6.0	6.0
Export crops		
Coffee	6.0	6.0
Tea	6.0	6.0
Other export crops	6.0	8.0
Livestock		
Cattle	6.0	6.0
Sheep and goats	6.0	6.0
Swine	6.0	6.0
Poultry	6.0	6.0
Raw milk	6.0	6.0
Eggs	6.0	6.0
Other livestock products	6.0	6.0
Forestry	6.0	6.0
Fishing	6.0	6.0
Mining	6.0	6.0
Food processing		
Meat, fish, and dairy products	3.0	3.0
Processed cereals	3.0	3.0
Processed coffee	2.0	6.0
Processed tea	2.0	6.0
Bakery and processed sugar	2.0	2.0
Traditional beverages	4.0	4.0
Modern beverages	4.0	4.0
Tobacco	4.0	4.0
Other manufacturing		
Textile and clothing	3.0	3.0
Wood, paper, and printing	4.0	4.0
Chemicals	4.0	4.0
Nonmetallic minerals	4.0	4.0
Furniture and other manufactured products	4.0	4.0
Electricity, gas, and water	4.0	4.0
Construction	4.0	4.0
Services		
Wholesale and retail trade	4.0	4.0
Hotels and restaurants	4.0	4.0
Transport	4.0	4.0
Communications	4.0	4.0
Finance and insurance	4.0	4.0
Real estate	4.0	4.0
Business services	4.0	4.0
Repair	4.0	4.0
Public administration	4.0	4.0
Education	4.0	4.0
Health	4.0	4.0
Other personal services	4.0	4.0

Source: Authors.
Note: Substitution elasticities for imports are those used in the Armington function; substitution elasticities for exports are those used in constant elasticity of transformation (CET) functions.

Table C.3 Sensitivity test: Production comparison

	Substitution elasticity 25% higher than in report			Substitution elasticity 25% lower than in report	
Subsector	Level of production, 2015	Difference from model result (%)	Model results in report	Level of production, 2015	Difference from model result (%)
Crops (thousand metric tons)					
Grains					
Wheat	46	0.6	45	45	-0.6
Maize	161	0.0	161	162	0.8
Rice	239	1.0	237	234	-1.4
Sorghum	264	0.0	264	264	0.0
Roots and tubers					
Irish potatoes	2,107	0.0	2,108	2,110	0.1
Sweetpotatoes	949	0.0	949	949	0.0
Cassava	1,060	0.0	1,060	1,060	0.0
Other roots	209	0.0	209	209	0.0
Other crops					
Pulses	432	0.0	432	432	0.0
Vegetables	912	0.0	912	912	0.0
Bananas	4,610	0.0	4,609	4,608	0.0
Fruits (excluding bananas)	598	0.0	598	598	0.0
Oilseeds	66	-0.1	66	67	0.1
Export crops					
Coffee	74	0.0	74	73	-0.7
Tea	56	0.4	56	55	-0.4
Other export crops	439	3.1	426	409	-4.0
Livestock (billion RWF)					
Cattle	28	-0.1	28	28	0.1
Sheep and goats	4	-0.1	4	4	0.1
Swine	3	-0.1	3	3	0.2
Poultry	4	-0.1	4	4	0.1
Raw milk	34	0.1	34	34	-0.1
Eggs	6	0.0	6	6	0.0
Other livestock	4	0.5	4	4	-0.6
Forestry (billion RWF)	82	0.0	82	83	0.1
Fishing (billion RWF)	12	0.1	11	11	-0.2

Source: Numbers are model simulation results derived from authors' calculations using 2006 data from MINAGRI (2007).

Notes: Substitution elasticities for imports are those used in the Armington function; substitution elasticities for exports are those used in constant elasticity of transformation (CET) functions. RWF, Rwandan franc.

Table C.4 Sensitivity test: Growth rate comparison (percent)

Indicator/ subsector	Substitution elasticity 25% higher than in report: Annual growth rate, 2015	Substitution elasticity 25% higher than in report: Difference from model result	Model results in report	Substitution elasticity 25% lower than in report: Annual growth rate, 2015	Substitution elasticity 25% lower than in report: Difference from model result
GDP	6.9	0.1	6.9	6.9	-0.1
AgGDP	6.5	0.1	6.5	6.5	-0.1
Cereals	8.0	0.4	7.9	7.9	-0.6
Roots	4.1	0.0	4.1	4.1	0.0
Other staples	5.6	0.0	5.6	5.6	-0.1
Export crops	8.7	0.2	8.6	8.6	-0.3
Livestock	12.3	0.0	12.3	12.3	0.0
Other agriculture	8.6	0.0	8.6	8.6	0.0
Industry	7.4	-0.3	7.5	7.5	0.4
Food processing	10.1	-0.4	10.2	10.2	0.5
Services	7.2	0.3	7.2	7.2	-0.3
Trade	8.3	0.4	8.3	8.3	-0.4
Transport	8.5	0.1	8.5	8.5	-0.1

Source: Numbers are model simulation results derived from authors' calculations using 2006 data from MINAGRI (2007).

Notes: Substitution elasticities for imports are those used in the Armington function; substitution elasticities for exports are those used in constant elasticity of transformation (CET) functions. AgGDP, agricultural gross domestic product; GDP, gross domestic product.

Table C.5 Sensitivity test: Poverty rate comparison (percent)

Region	Substitution elasticity 25% higher than in report: Poverty rate, 2015	Substitution elasticity 25% higher than in report: Difference from model result	Model results in report	Substitution elasticity 25% lower than in report: Poverty rate, 2015	Substitution elasticity 25% lower than in report: Difference from model result
National	36.86	-0.1	36.90	36.98	0.2
Urban	18.52	0.6	18.41	18.41	0.0
Rural	40.52	-0.2	40.59	40.69	0.2

Source: Numbers are model simulation results derived from authors' calculations using 2006 data from MINAGRI (2007).

Note: Substitution elasticities for imports are those used in the Armington function; substitution elasticities for exports are those used in constant elasticity of transformation (CET) functions.

References

African Union. 2003. Declaration on agriculture and food security in Africa. Assembly/AU/Decl.7 (II). Declaration adopted at the Assembly of the African Union, Second Ordinary Session, July 10–12, Maputo, Mozambique. http://www.africa-union.org/Official_documents/Decisions_Declarations/Assembly%20final/Decl%20on%20the%20Agriculture%20and%20Food%20Security%20Maputo%202003.pdf. Accessed May 20, 2010.

Akita, T., and M. Kawamura. 2002. Regional income inequality in China and Indonesia: A comparative analysis. Paper presented at the 42nd Congress of the European Regional Science Association, August 27–31, Dortmund, Germany.

Alliance of CGIAR Centers. 2008. *Alliance of CGIAR Centers' best bets to boost crop yields in Sub-Saharan Africa: Current products of the Alliance of CGIAR Centers Research*. Rome.

Angelsen, A., and D. Kaimowitz. 1999. Rethinking the causes of deforestation: Lessons from economic models. *World Bank Research Observer* 14 (1): 73–98.

Ansoms, A. 2008. *A Green Revolution for Rwanda? The political economy of poverty and agrarian change*. Discussion Paper 2008.06. Antwerp, Belgium: Institute of Development Policy and Management, University of Antwerp.

Arnault Emini, C. 2007. The 2006 social accounting matrix of Rwanda: Methodology note. November. Yaounde, Cameroon, University of Yaounde II. Unpublished memo.

Bezemer, D., and D. Headey. 2008. Agriculture, development and urban bias. *World Development* 34 (8): 1342–1364.

Bigsten, A., and A. Isaksson. 2008. *Growth and poverty in Rwanda: Evaluating the EDPRS 2008-12*. Country Economic Report 2008:3. Department for Policy and Methodology, Swedish International Development Cooperation Agency. Stockholm, Sweden: Edita Communication. http://www.hgu.gu.se/Files/nationalekonomi/Personal/Isaksson%20Ann-Sofie/SIDA%20report%202008.pdf.

Blarel, B., P. Hazell, F. Place, and J. Quiggin. 1992. The economics of farm fragmentation: Evidence from Ghana and Rwanda. *World Bank Economic Review* 6 (2): 233–254.

Breisinger, C., X. Diao, and J. Thurlow. 2009. Modeling growth options and structural change to reach middle-income country status: The case of Ghana. *Economic Modeling* 26 (2): 514–525.

Byerlee, D., X. Diao, and C. Jackson. 2005. *Agriculture, rural development, and pro-poor growth: Country experiences in the post-reform era*. Synthesis paper for the Operationalizing Pro-Poor Growth in the 1990s Project. Washington, D.C.: World Bank.

Chen, S., and M. Ravallion. 2000. *How did the world's poorest fare in the 1990s?* Policy Research Working Paper 2409. Washington, D.C.: World Bank.

Christiaensen, L., L. Demery, and J. Kühl. 2006. *The role of agriculture in poverty reduction: An empirical perspective.* Policy Research Working Paper 4013. Washington, D.C.: World Bank.

Clay, D. C. 1996. *Fighting an uphill battle: Population pressure and declining land productivity in Rwanda.* International Development Working Paper 58. East Lansing, Mich., U.S.A.: Department of Agricultural Economics and Department of Economics, Michigan State University.

Clay, D. C., T. Reardon, and J. Kangasniemi. 1998. Sustainable intensification in the highland tropics: Rwandan farmers' investments in land conservation and soil fertility. *Economic Development and Cultural Change* 46 (2): 351-377.

Clay, D. C., F. Byiringiro, J. Kangasniemi, T. Reardon, B. Sibomana, L. Uwamariya, and D. Tardif-Douglin. 1995. *Promoting food security in Rwanda through sustainable agricultural productivity: Meeting the challenges of population pressure, land degradation and poverty.* International Development Paper 17. East Lansing, Mich., U.S.A.: Department of Agricultural Economics and Department of Economics, Michigan State University.

Collier, P. 2002. Primary commodity dependence and Africa's future. In *Annual World Bank Conference on Development Economics proceedings.* Washington, D.C.: World Bank.

Diao, X., and P. Dorosh. 2007. Demand constraints on agricultural growth in East and Southern Africa: A general equilibrium analysis. *Development Policy Review* 25 (3): 275-292.

Diao, X., J. Rattsø, and H. E. Stokke. 2005. International spillovers, productivity growth and openness in Thailand: An intertemporal general equilibrium model analysis. *Journal of Development Economics* 76 (2): 429-450.

Diao, X., E. Yeldan, and T. Roe. 1998. A simple dynamic applied general equilibrium model of a small open economy: Transitional dynamics and trade policy. *Journal of Economic Development* 23 (1): 77-101.

Diao, X., P. Hazell, D. Resnick, and J. Thurlow. 2007. *The role of agriculture in development: Implications for Sub-Saharan Africa.* Research Report 153. Washington, D.C.: International Food Policy Research Institute.

Ellis, F. 2005. Small farms, livelihood diversification, and rural-urban transitions: Strategic issues in Sub-Saharan Africa. In *The future of small farms: Proceedings of a research workshop, Wye, U.K., June 26-29, 2005.* Washington, D.C.: International Food Policy Research Institute. http://www.ifpri.org/sites/default/files/publications/sfproc.pdf. Accessed January 6, 2010.

Fan, S., B. Yu, and A. Saurkar. 2008. Public spending in developing countries: Trends, determination and impact. In *Public expenditures, growth and poverty: Lessons from developing countries,* ed. S. Fan. Baltimore: Johns Hopkins University Press.

Fleskens, L. 2007. *Prioritizing rural public works interventions in support of agricultural intensification.* Annex report. Helpage, Rwanda: Catalyze Accelerated Agricultural Intensification for Social and Environmental Stability, International Fertilizer Development Center.

Fowler, M., C. Gasirabo, S. Kanyarukiga, and A. Mutijima. 2007. *Rwanda public expenditure review—Agriculture.* Report for the Ministry of Agriculture and Animal

Resources, Rwanda. http://www.amis.minagri.gov.rw/sites/default/files/user/PER_final_version_030507_revised_by_Sam1.pdf. Accessed May 2009.

Gohin, A. 2005. The specification of price and income elasticities in computable general equilibrium models: An application of latent separability. *Economic Modeling* 22 (5): 905–925.

Haggblade, S., P. Hazell, and T. Reardon. 2007. Strategies for stimulating equitable growth in the rural nonfarm economy. In *Transforming the rural nonfarm economy*, ed. S. Haggblade, P. Hazell, and T. Reardon. Baltimore: Johns Hopkins University Press.

Hazell, P., and X. Diao. 2005. The role of agriculture and small farms in economic development. In *The future of small farms: Proceedings of a research workshop, Wye, U.K., June 26–29, 2005*. Washington, D.C.: International Food Policy Research Institute. http://www.ifpri.org/sites/default/files/publications/sfproc.pdf. Accessed January 6, 2010.

ICARRD (International Conference on Agrarian Reform and Rural Development). 2006. A case study on the implications of the ongoing land reform on sustainable rural development and poverty reduction in Rwanda and the outcome report of the thematic dialogue held on 20th January 2006, January 20, Kigali, Rwanda. http://www.icarrd.org/en/icard_doc_down/case_Rwanda.pdf. Accessed May 2009.

IMF (International Monetary Fund). 2004. *Rwanda: Selected issues and statistical appendix*. IMF Country Report 04/383. Washington, D.C.

Ivanic, M., and W. Martin. 2008. *Implications of higher global food prices for poverty in low-income countries*. Policy Research Working Paper 4594. Washington, D.C.: World Bank.

Johnson, M., P. Hazell, and A. Gulati. 2003. The role of intermediate factors markets in Asia's Green Revolution: Lessons for Africa? *American Journal of Agricultural Economics* 85 (5): 1211–1216.

Kanbur, R., and X. Zhang. 2005. Fifty years of regional inequality in China: A journey through central planning, reform, and openness. *Review of Development Economics* 9 (1): 87–106.

Kelly, V., E. Mpyisi, E. Shingiro, and J. B. Nyarwaya. 2001a. *Agricultural intensification in Rwanda: An elusive goal. Fertilizer use and conservation investment*. International Development Collaborative Policy Brief RW-FSRP-PB-01. East Lansing, Mich., U.S.A.: Department of Agricultural Economics, Michigan State University.

Kelly, V., E. Mpyisi, A. Murekezi, and D. Neven. 2001b. Fertilizer consumption in Rwanda: Past trends, future potential, and determinants. Paper prepared for Policy Workshop on Fertilizer Use and Marketing, February 22–23, Kigali, Rwanda.

King, R. P., and D. Byerlee. 1978. Factor intensities and locational linkages of rural consumption patterns in Sierra Leone. *American Journal of Agricultural Economics* 60 (2): 197–206.

Lofgren, H., R. Harris, and S. Robinson. 2001. *A standard computable general equilibrium (CGE) model in GAMS*. Trade and Macroeconomics Discussion Paper 75. Washington, D.C.: International Food Policy Research Institute.

Lopez, H., and Q. Wodon. 2005. The economic impact of armed conflict in Rwanda. *Journal of African Economies* 14 (4): 586–602.

Lopez, R., and M. Niklitschek. 1991. Dual economic growth in poor tropical areas. *Journal of Development Economics* 36 (2): 189–211.

Maxwell, S., and R. Slater. 2003. Food policy old and new. *Development Policy Review* 21 (5–6): 531–553.

MINAGRI (Ministry of Agriculture and Animal Resources). 2004a. *National agricultural policy*. Kigali, Rwanda.

——. 2004b. *Strategic plan for agricultural transformation in Rwanda. Main document*. Kigali, Rwanda.

——. 2007. Unpublished statistics data files. Kigali, Rwanda.

——. 2008. *Total cost of EDPRS agriculture sector*. Kigali, Rwanda.

MINECOFIN (Ministry of Finance and Economic Planning). 2000. *Vision 2020*. Kigali, Rwanda.

——. 2002. *Poverty reduction strategy paper*. Kigali, Rwanda.

——. 2003. *Household living conditions survey, 1999/2001*. Kigali, Rwanda.

——. 2006. *Annual finance laws 1999-2006*. Kigali, Rwanda.

——. 2007a. *Household living conditions survey, 2005/2006*. Kigali, Rwanda.

——. 2007b. *Rwanda agricultural survey 2006*. Kigali, Rwanda: National Institute of Statistics of Rwanda and MINECOFIN, Republic of Rwanda.

——. 2008. *Economic development and poverty reduction strategy 2008-12*. Kigali, Rwanda.

Morris, M., L. Drake, K. Ezemenari, and X. Diao. 2008. *Promoting sustainable pro-poor growth in Rwandan agriculture: What are the policy options?* Africa Region Working Paper 114. Washington, D.C.: World Bank.

National Institute of Statistics. 2006. *Preliminary poverty update report: Integrated living conditions survey 2005/06*. Kigali, Rwanda.

Ndulu, B., and S. O'Connell. 2006. Policy plus: African growth performance 1960–2000. In *The political economy of economic growth in Africa 1960-2000*, ed. B. Ndulu, S. O'Connell, R. Bates, P. Collier, and C. Soludo. Cambridge: Cambridge University Press.

Nkeshimana, G. 2008. The relation between land management practices, soil physical properties and erosion in coffee systems in south Rwanda. M.S. thesis, Wageningen University, Wageningen, Netherlands.

OECD-FAO (Organisation for Economic Co-operation and Development–Food and Agriculture Organization of the United Nations). 2007. *Agricultural outlook 2007-16*. Rome and Paris. http://www.oecd.org/dataoecd/6/10/38893266.pdf. Accessed June 2008.

REMA (Rwanda Environment Management Authority). 2009. *Rwanda state of environment and outlook report*. Kigali, Rwanda.

Roose, E., and F. Ndayizigiye. 1997. Agroforestry, water and soil fertility management to fight erosion in the tropical mountains of Rwanda. *Soil Technology* 11 (1): 109–119.

Thirtle, C., L. Lin, and J. Piesse. 2003. The impact of research-led agricultural productivity growth on poverty reduction in Africa, Asia and Latin America. *World Development* 31 (12): 1959–1975.

Thurlow, J. 2004. Trade liberalization and poverty in South Africa: A computable general equilibrium and microsimulation analysis of past and potential trade policies. Ph.D. dissertation, University of Natal, Durban, South Africa.

USDA (United States Department of Agriculture). 2008. *USDA agricultural projections to 2017; Long-term projections report OCE-2008-1*. Washington, D.C.

Verpoorten, M., and L. Berlage. 2004. Genocide and land scarcity: Can Rwandan rural households manage? Paper presented at African Development and Poverty Reduction: The Macro-Micro Linkage Forum, October 13–15, Somerset West, South Africa. http://www.tips.org.za/files/Genocide_and_Land_Scarcity_Verpooten.pdf. Accessed January 6, 2010.

Verwimp, P. 2003. Testing the double-genocide thesis for central and southern Rwanda. *Journal of Conflict Resolution* 47 (4): 423–442.

von Braun, J., with A. Ahmed, K. Asenso-Okyere, S. Fan, A. Gulati, J. Hoddinott, R. Pandya-Lorch, M. W. Rosegrant, M. Ruel, M. Torero, T. van Rheenen, and K. von Grebmer. 2008. *High food prices: The what, who, and how of proposed policy actions*. Policy Brief 1A. Washington, D.C.: International Food Policy Research Institute.

von Braun, J., H. de Haen, and J. Blanken. 1991. *Commercialization of agriculture under population pressure: Effects on production, consumption, and nutrition in Rwanda*. Research Report 85. Washington, D.C.: International Food Policy Research Institute.

Wilhelmsson, M. 2002. Household expenditure patterns for housing attributes: A linear expenditure system with hedonic prices. *Journal of Housing Economics* 11 (1): 75–93.

World Bank. 2007. *Rwanda country economic memorandum (CEM)*. Washington, D.C.

———. 2008a. *World development indicators*. Washington, D.C. CD-ROM.

———. 2008b. *World development report 2008: Agriculture for development*. Washington, D.C.

Yu, W., T. Hertel, P. V. Prechel, and J. S. Eales. 2003. Projecting world food demand using alternative demand systems. *Economic Modeling* 21 (1): 99–129.

About the Authors

Xinshen Diao is a senior research fellow in the Development Strategy and Governance Division of the International Food Policy Research Institute.

Shenggen Fan is director general of the International Food Policy Research Institute. At the time this study was conducted, he was head of the Development Strategy and Governance Division of the Institute.

Sam Kanyarukiga is president of Management and Advisory Services for Development, a nongovernmental organization that specializes in agricultural research and monitoring and evaluation of rural cooperatives in Rwanda.

Bingxin Yu is a postdoctoral fellow in the Development Strategy and Governance Division of the International Food Policy Research Institute.

Index

Page numbers for entries occurring in figures are suffixed by *f*, those for entries occurring in notes by *n*, and those for entries in tables by *t*.

Made in the USA
Monee, IL
04 November 2026

67379773R00079